DYNAMICS—THE GEOMETRY OF BEHAVIOR

PART 1: PERIODIC BEHAVIOR

VISMATH: THE VISUAL MATHEMATICS LIBRARY

VOLUME 1

THE VISUAL MATHEMATICS LIBRARY
Ralph H. Abraham, Editor

VISMATH BOOKS:

VOLUME 0: Abraham and Shaw, MANIFOLDS AND MAPPINGS

VOLUME 1: Abraham and Shaw, DYNAMICS—THE GEOMETRY OF BEHAVIOR
Part 1: Periodic Behavior

VOLUME 2: Abraham and Shaw, DYNAMICS—THE GEOMETRY OF BEHAVIOR
Part 2, Chaotic Behavior

VOLUME 3: Abraham and Shaw, DYNAMICS— THE GEOMETRY OF BEHAVIOR
Part 3: Global Behavior

VOLUME 4: Abraham & Shaw, DYNAMICS— THE GEOMETRY OF BEHAVIOR
Part 4: Bifurcation Behavior

VISMATH COMPUTER GRAPHIC PROGRAMS:

DISK 1: Abraham and Norskog, PHASE PORTRAITS IN THE PLANE

DISK 2: Abraham and Norskog, CHAOTIC ATTRACTORS IN 3D

DISK 3: Abraham and Norskog, BIFURCATION EXPERIMENTS

VISMATH LIBRARY FILM SERIES

THE LORENZ SYSTEM: Bruce Stewart, 25 minutes, color 16mm.

CHAOTIC CHEMISTRY: Robert Shaw, Jean-Claude Roux and Harry Swinney, 20 minutes,
black & white, 16 mm.

CHAOTIC ATTRACTORS OF DRIVEN OSCILLATORS : J. P. Crutchfield, 12 minutes,
black & white, 16 mm.

THE SCIENCE FRONTIER EXPRESS SERIES

THE DRIPPING FAUCET AS A MODEL CHAOTIC SYSTEM, Robert Shaw
ON MORPHODYNAMICS, Selected papers by Ralph Abraham
COMPLEX DYNAMICAL SYSTEMS, Selected papers by Ralph Abraham

SERIES FOREWORD

During the Renaissance, algebra was resumed from Near Eastern sources, and geometry from the Greek. Scholars of the time became familiar with classical mathematics. When calculus was born in 1665, the new ideas spread quickly through the intellectual circles of Europe. Our history shows the importance of the diffusion of these mathematical ideas, and their effects upon the subsequent development of the sciences and technology.

Today, there is a cultural resistance to mathematical ideas. Due to the widespread impression that mathematics is difficult to understand, or to a structural flaw in our educational system, or perhaps to other mechanisms, mathematics has become an esoteric subject. Intellectuals of all sorts now carry on their discourse in nearly total ignorance of mathematical ideas. We cannot help thinking that this is a critical situation, as we hold the view that mathematical ideas are essential for the future evolution of our society.

The absence of visual representations in the curriculum may be part of the problem, contributing to mathematical illiteracy, and to the math-avoidance reflex. This series is based on the idea that mathematical concepts may be communicated easily in a format which combines visual, verbal, and symbolic representations in tight coordination. It aims to attack math ignorance with an abundance of visual representations.

In sum, the purpose of this series is to encourage the diffusion of mathematical ideas, by presenting them *visually*.

THE VISUAL MATHEMATICS LIBRARY: VISMATH VOLUME 1

DYNAMICS—THE GEOMETRY OF BEHAVIOR

Part 1: Periodic Behavior

With 342 illustrations

by

Ralph H. Abraham

and

Christopher D. Shaw

University of California
Santa Cruz, CA 95064

Aerial Press, Inc.
P.O.Box 1360, Santa Cruz, California, 95061

Library of Congress Cataloging in Publication Data

Library of Congress Catalog Card Number: 81-71616
ISBN 0-942344-01-4 (Volume 1)
ISBN 0-942344-00-6 (4 volume set)

Second Printing: March, 1984
Third Printing: March, 1985

Original text reproduced by Aerial Press, Inc., from camera-ready copy prepared by the authors.
Copyright © by Aerial Press, Inc.

Printed in the United States of America.

CONTENTS

CONTENTS

PREFACE

what is dynamics?

Dynamics is a field emerging somewhere between mathematics and the sciences. In our view, it is the most exciting event on the concept horizon for many years. The new concepts appearing in dynamics extend the conceptual power of our civilization, and provide new understanding in many fields.

the visual math format

We discovered, while working together on the illustrations for a book(1)[*] in 1978, that we could explain mathematical ideas visually, within an easy and pleasant working partnership. In 1980, we wrote an expository article on dynamics and bifurcations[1][**], using hand-animation to emulate the *dynamic picture technique* universally used by mathematicians in talking among themselves: a picture is drawn slowly, line-by-line, along with a spoken narrative — the dynamic picture and the narrative tightly coordinated.

Our efforts inevitably exploded into four volumes of this series, of which this is the first. The dynamic picture technique, evolved through our work together, and in five years of computer graphic experience with the *Visual Math Project* at the University of California at Santa Cruz, is the basis of this work. About two-thirds of the book is devoted to visual representations, in which four colors are used according to a strict code.

Moving versions of the phase portraits, *actual dynamic pictures,* will be made available as *computer graphic programs* on floppy disks for home computers, and as *videotapes, videodiscs, or films.*

on the suppression of symbols

Math symbols have been kept to a minimum. In fact, they are almost completely suppressed. Our purpose is to make the book work for readers who are not practiced in symbolic representations. We rely exclusively on visual representations, with brief verbal explanations. Some formulas are shown with the applications, as part of the graphics, but are not essential. However, *this strategy is exclusively pedagogic.* We do not want anyone to think that we consider symbolic representations unimportant in mathematics. On the contrary, this field evolved primarily in the symbolic realm throughout the classical period. Even now, a full understanding of our subject demands a full measure of formulas, logical expressions, and technical intricacies from all branches of mathematics. A brief introduction to these is included in the *appendix.*

[*]Numbers in parentheses refer to the Bibliography, and usually suggest further introductory reading.
[**]Numbers in brackets refer to the Notes, and usually suggest further advanced reading.

our goals

We have created this book as a short-cut to the research frontier of dynamical systems: theory, experiments, and applications. It is our goal—we know we may fail to reach it—to provide any interested person with an acquaintance with the basic concepts:

   *state spaces: manifolds—geometric models for the virtual states of a system

   *attractors: static, periodic, and chaotic—geometric models for the asymptotic behavior of a system

   *separatrices: repellors, saddles, insets, tangles—defining the boundaries of regions (basins) dominated by different behaviors.

The ideas included are selected from the classical literature of dynamics: covering the period from 1600 to 1950. The sequel, Volume 2 of this series, is devoted to the recent developments: 1950 to the present, belonging to the tradition begun by Poincaré.

prerequisite background

We assume nothing in the way of prior mathematical training, beyond vectors in three dimensions, and complex numbers. Nevertheless, it will be tough going without a basic understanding of the simplest concepts of calculus. All the essential ideas will be presented in Volume 0 of this series.

acknowledgements

Our first attempt at the pictorial style used here evolved from the helpful comments of Len Fellman, Alan Garfinkel, Norman Packard, Steve Smale, Joel Smoller, and Chris Zeeman on the first draft of *Dynamics, a Visual Introduction [1],* and the encouragement of the editor, Gene Yates, during the Summer of 1980. Our next effort, the preliminary draft of Volume 2 of this series, was circulated among friends in the Summer of 1981. Extensive feedback from them has been very influential in the evolution of this volume, and the whole series. In particular, the suggestions from Michael Arbib led to the creation of this volume, in preparation for Volume 2. The responses from Alan Garfinkel were responsible for our heavy emphasis on coupled oscillators and entrainment, while those from Tim Poston and Jim Crutchfield were very influential throughout the text. A splendidly illustrated letter from George Francis caused substantial improvements in our visual representations. And for many helpful comments, we are also grateful to:

Ethan Akin	John Guckenheimer	Jean-Michel Kantor	Charles Muses	Jim Swift
Larry Cuba	Phil Holmes	Bob Lansdon	Lee Rudolph	Bob Williams
Richard Cushman	Dan Joseph	Jim McGill	Mike Shub	

In contrast to Volume 2, the first draft of this volume has not had the benefit of extensive circulation. So we are especially grateful to Alan Garfinkel and Tim Poston for their careful reading and extensive comments.

Thanks are also due to Isaac Rabinowitz and Lance Norskog for expert guidance on text-processing, Nina Graboi for proofreading this volume, and to Claire Moore of Aerial Press for producing it. The generosity and good will of many dynamicists has been crucial in the preparation of this book; we thank them all. Finally, it is a pleasure to thank the National Science Foundation for financial support.

Ralph H. Abraham   Christopher D. Shaw   Santa Cruz, California   March, 1982

dedicated to
Lord Rayleigh

DYNAMICS—THE GEOMETRY OF BEHAVIOR

Part 1: Periodic Behavior

DYNAMICS HALL OF FAME

Dynamics has evolved into three disciplines: applied, mathematical, and experimental. Applied dynamics is the oldest. Originally regarded as a branch of natural philosophy, or physics, it goes back to Galileo at least. It deals with the concepts of change, rate of change, rate of rate of change, and so on, as they occur in various natural phenomena. We take these concepts for granted, but they emerged into our conciousness only in the fourteenth century[1].

Mathematical dynamics begins with Newton, and has become a large and active branch of pure mathematics. This includes the theory of ordinary differential equations, now a classical subject. But since Poincaré, the newer methods of topology and geometry have dominated the field.

Experimental dynamics is an increasingly important branch of the subject. Founded by Galileo, it showed little activity until Rayleigh, Duffing, and Van der Pol. Experimental techniques have been revolutionized with each new development of technology. Analog and digital computers are now in vogue. They are accelerating the advance of the research frontier, making experimental work more significant than ever.

This chapter presents a few words of description for some of the leading figures of the history of dynamics. Their positions in a two-dimensional tableau — date versus specialty (applied, mathematical, or experimental dynamics) — are shown in Table 1. Those included are not more important than numerous others, but limitations of space and knowledge prevent us from giving a more complete museum here.

TABLE 1. THE HISTORY OF DYNAMICS			
Date	APPLIED DYNAMICS	MATHEMATICAL DYNAMICS	EXPERIMENTAL DYNAMICS
1600	Kepler		Galileo
1650	Huyghens	Newton Leibniz	
1700			
1750		Euler	
		Lagrange	
1800			
1850	Helmholtz Rayleigh	Poincaré Lie Liapounov	Rayleigh
1900			Duffing van der Pol
1950	Lotka Volterra Rashevsky	Birkhoff Andronov Cartwright	Hayashi

Galileo Galilei, 1564-1642. One of the first to deal thoroughly with the concept of acceleration, Galileo founded dynamics as a branch of natural philosophy. The close interplay of theory and experiment, characteristic of this subject, was founded by him.

photo courtesy of D. J. Struik, A Concise History of Mathematics, Dover Publications, New York (1948)

Johannes Kepler, 1571-1630. The outstanding and original exponent of applied dynamics, Kepler made use of extensive interaction between theory and observation, to understand the planetary motions.

photo courtesy of Kepler, Gesammelte Werke. Beck, München (1960)

Isaac Newton, 1642-1727. Mathematical dynamics, as well as the calculus on which it is based, was founded by Newton at age 23. Applications and experiments were basic to his ideas, which were dominated by the doctrine of determinism. His methods were geometric.

photo courtesy of the Trustees of the British Museum

Gottfried Wilhelm Leibniz, 1646-1716. The concepts of calculus, mathematical dynamics, and their implications for natural philosophy, occurred independently to Leibniz. His methods were more symbolic than geometric.

photo courtesy of the Trustees of the British Museum

Leonhard Euler, 1707-1783. Primarily known for his voluminous contributions to algebra, Euler developed the techniques of analysis which were to dominate mathematical dynamics throughout its classical period.

photo courtesy of E. T. Bell, Men of Mathematics, Simon and Schuster, New York (1937)

Joseph-Louis Lagrange, 1736-1813. A disciple of Euler, Lagrange developed the analytical method to extremes, and boasted that his definitive text on the subject contained not a single illustration.

photo courtesy of the Bibliotèque Nationale, Paris, France.

Marius Sophus Lie, 1842-1899. In combining the ideas of symmetry and dynamics, Lie built the foundations for a far-reaching extension of dynamics, the theory of groups of transformations.

photo courtesy of Minkowski, H., Briefe an David Hilbert, Mit Beiträgen und herausgegeben von L. Rüdenberg, H. Zassenhaus; Springer-Verlag, Heidelberg (1973)

John William Strutt,   Baron Rayleigh, 1842-1919. In a career of exceptional length and breadth, spanning applied mathematics, physics, and chemistry, Rayleigh dwelled at length on acoustical physics. In this context, he revived the experimental tradition of Galileo in dynamics, laying the foundations for the theory of nonlinear oscillations. His text on acoustics, published in 1877, remains to this day the best account of this subject[2].

photo courtesy of Applied Mech. Rvws. 26 (1973)

Jules Henri Poincaré, 1854-1912. Known for his contributions to many branches of pure mathematics, Poincaré devoted the majority of his efforts to mathematical dynamics. Among the first to accept the fact that the classical analytical methods of Euler and Lagrange had serious limitations, he revived geometrical methods. The results were revolutionary for dynamics, and gave birth to topology and global analysis as well. These branches of pure mathematics are very active yet.

photo courtesy of the Library of Congress, Washington, D.C.. U.S.A.

Aleksandr Mikhailovich Liapounov, 1857-1918. Another pioneer of geometric methods in mathematical dynamics, Liapounov contributed basic ideas of stability.

photo courtesy of Akademija Nauk, SSSR (1954)

Georg Duffing, 1861-19??. A serious experimentalist, Duffing studied mechanical devices to discover geometric properties of dynamical systems. The theory of oscillations was his explicit goal.

Photo not available —
Please communicate with us if
you have leads on photo
or illustration.

George David Birkhoff, 1884-1944. The first dynamicist in the New World, Birkhoff picked up where Poincaré left off. Although a geometer at heart, he discovered new symbolic methods. He saw beyond the theory of oscillations, created a rigorous theory of ergodic behavior, and foresaw dynamical models for chaos.

photo courtesy of G. D. Birkhoff, Collected Mathematical Papers, American Mathematical Society, New York (1950)

Balthasar van der Pol,   1889-1959.
The first radio transmitter became, in
the hands of this outstanding experimentalist, a high-speed laboratory of
dynamics. Many of the basic ideas of
modern experimental dynamics came
out of this laboratory.

*photo courtesy of Balthasar van der Pol, Selected
Scientific Papers, Vol. 1,   H. Bremmer and C.J.
Boukamp (eds.), North-Holland, Amsterdam (1960)*

Nicholas   Rashevsky,   1899-1972.
From antiquity until the 1920's, applied
dynamics meant physics. At last, the
important applications to the biological
and social sciences came into view, in
the visionary minds of the general scien-
tists, Lotka, Volterra, and Rashevsky.

photo courtesy of Bull. Math. Biophys. 34 (1972)

Mary Lucy Cartwright, 1900- Dame Cartwright, together with J. E. Littlewood, revived dynamics in England, during World War II. Inspired by the work of Van der Pol, they obtained important results on the ultraharmonics of forced electronic oscillations, using analytical and topological methods.

photo courtesy of Math. Gazette 36 (1952)

Chihiro Hayashi, 1911. The experiments of dynamicists were restricted to a few simple systems (Duffing's system, Van der Pol's system, etc.) until the appearance of the general purpose analog computer. One of the creators of this type of machine, and the first to fully exploit one as a laboratory of dynamics, Hayashi contributed much to our knowledge of oscillations.

photo courtesy of Ch. Hayashi, Selected Papers on Nonlinear Oscillators, Kyoto (1975)

BASIC CONCEPTS OF DYNAMICS

The key to the geometric theory of dynamical systems created by Poincaré, is the *phase portrait* of a dynamical system. The first step in drawing this portrait is the creation of a geometric model for the set of all possible states of the system. This is called the *state space*. On this geometric model, the dynamics determine a cellular structure of *basins* enclosed by *separatrices*. Within each cell, or basin, is a nucleus called the *attractor*. The states which will actually be observed in this system are the attractors. Thus, the portrait of the dynamical system, showing the basins and attractors, is of primary importance in applications. This chapter introduces these basic concepts.

1.1 STATE SPACES

The strategies for making mathematical models for observed phenomena have been evolving since ancient times. An organism--physical, biological, or social--is observed in different states. This *observed system* is the target of the modeling activity. Its states cannot really be described by only a few observable parameters, but we pretend that they can. This is the first step in the process of "mathematical idealization" and leads to a geometric model for the set of all idealized states: the *state space* of the model. Different models may begin with different state spaces. The relationship between the actual states of the real organism and the points of the geometric model is a fiction maintained for the sake of discussion, theory, thought, and so on: this is known as the *conventional interpretation*. This section describes some examples of this modeling process.

The simplest scheme is the one-parameter model. The early history of science used this scheme extensively.

1.1.1. The actual state of this waffle iron cannot be described completely by a single observable parameter, such as the temperature. But usually we find it convenient to pretend that it can. This pretense is an agreement, the *conventional interpretation,* within the modeling process. It is justified by its usefulness in describing the behavior of the device.

1.1.2. The correlation between the internal state of a complex system, such as a mammal, and a single observed parameter may be very good or very bad, depending on the context. In the case of George Washington, the oral temperature correlates better with his health than with his honesty.

TEMPERATURE

1.1.3. In these examples, the geometric model for the set of all (mathematically idealized) states is the real number line. This is one of the simplest state spaces.

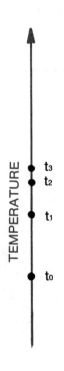

1.1.4. Observing the parameter for a while, it will probably change. The different values observed may be labeled by the time of their observation: the states observed at four different times are shown here.

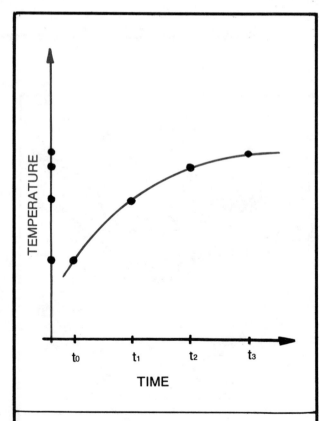

1.1.5. These data comprise a *time series* of observations, and are shown here as a graph. The vertical line represents the state space, the horizontal axis indicates time.

Closer observation may suggest two parameters for the description of a given state of the actual organism.

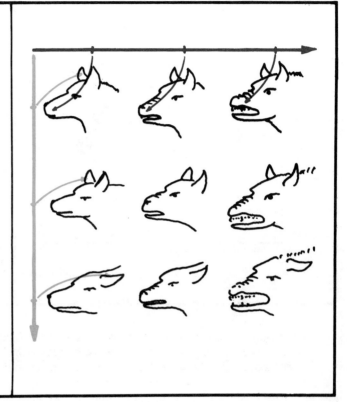

1.1.6. In this modeling scheme of Konrad Lorenz and Christopher Zeeman (1), two parameters are used for the emotional state of a dog. The two observed parameters are *ear attitude*, which correlates with the emotional state of fear, and *fang exposure*, corresponding to the degree of rage.

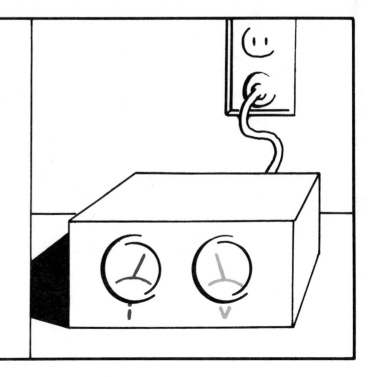

1.1.7. Electronic devices are simple to model, as the observations are explicitly numerical. This electronic 'black box' is provided with panel meters, which indicate the instantaneous values of voltage and current at specific points of the electronic network within the box.

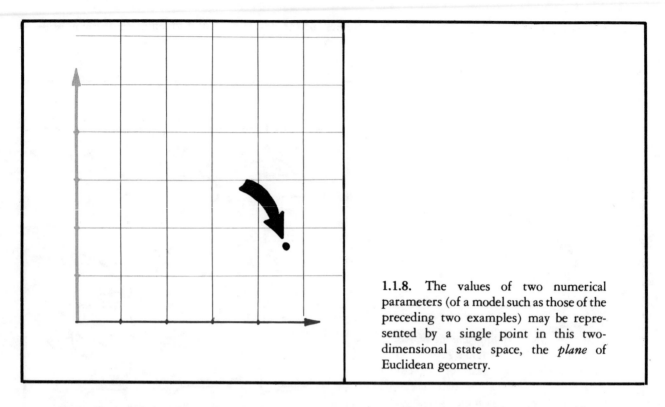

1.1.8. The values of two numerical parameters (of a model such as those of the preceding two examples) may be represented by a single point in this two-dimensional state space, the *plane* of Euclidean geometry.

Changes in the actual state of the system are observed, and are represented as a curve in the state space. Each point on this curve carries (implicitly at least) a label recording the time of observation. This is called a *trajectory* of the model.

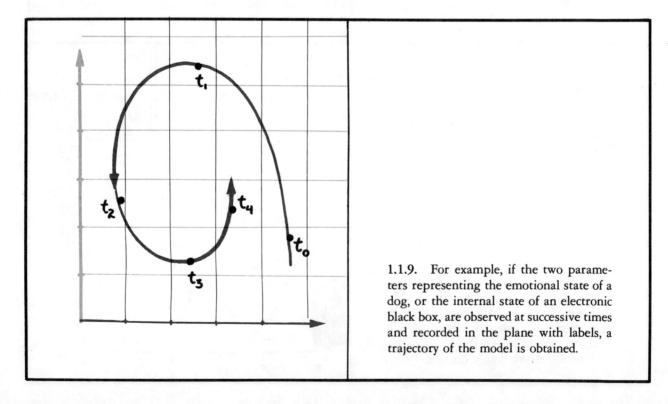

1.1.9. For example, if the two parameters representing the emotional state of a dog, or the internal state of an electronic black box, are observed at successive times and recorded in the plane with labels, a trajectory of the model is obtained.

Another style of representing the changing data is by its *time series,* which means the *graph* of a trajectory. We have already seen a time series, in a one-dimensional context. But this style of data representation may also be used in higher dimensions.

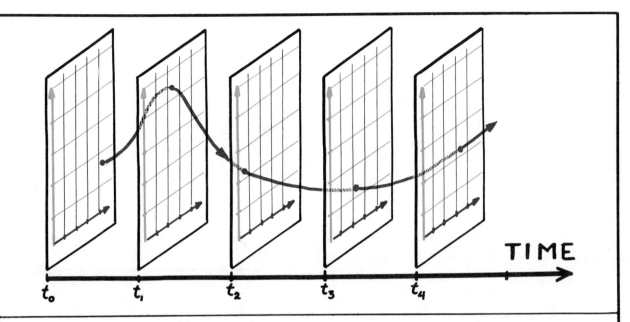

1.1.10. Here the vertical plane represents the state space, and the horizontal axis represents the time of observation. The parameters observed at a given time are plotted in the vertical plane passing through the appropriate point on the time axis.

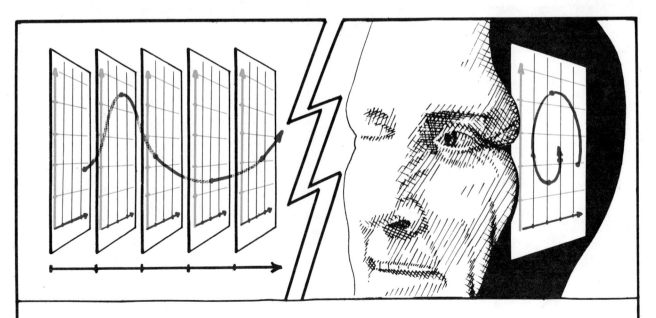

1.1.11. The trajectory may by obtained from the time series, by simply viewing it from the right angle— straight down the time axis from the end, infinitely far away.

Observing more parameters leads to models of higher dimensions.

1.1.12. Suppose that a 7 a.m., this athlete observes three of his body parameters (say: temperature, blood pressure, and pulse rate), records these three data as a point in three-dimensional space, and labels this point with the time of observation. This is a simple example of a three-dimensional state space.

Many phenomena require geometric models which are not simply coordinate spaces. In dynamical systems theory, the geometric models which are used are *manifolds*.

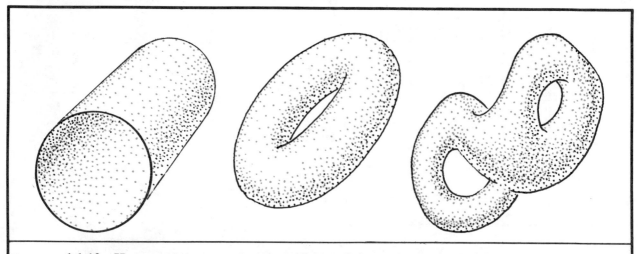

1.1.13. Here are some examples of manifolds. Other examples will arise in the sequel. They are made of pieces of flat spaces, bent, and glued together.

More about manifolds may be found in Volume 0 of this series.

1.2   DYNAMICAL SYSTEMS

At this point, the history of a real system has been represented graphically, as a *trajectory* in a geometric *state space*. An alternative representation is the *time series,* or *graph,* of the trajectory. The dynamical concepts of the middle ages included these kinds of representation. But in the 1660's, something new was added — *the instantaneous velocity, or derivative, of vector calculus* — by Newton. As dynamical systems theory evolved, the *velocity vectorfield* emerged as one of the basic concepts. Trajectories determine velocity vectors, by the *differentiation* process of Calculus. Conversely, velocity vectors determine trajectories, by the *integration* process of Calculus.

This is the differentiation process, which determines the velocity vectorfield from the trajectories.

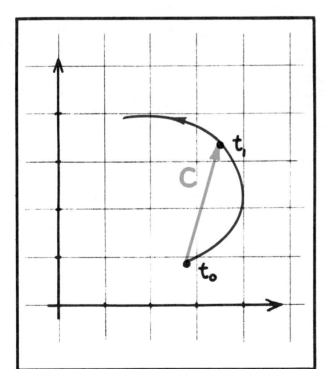

1.2.1. On this trajectory, the states observed at two different times, t_0 and t_1, are connected by a *bound vector,* represented here by a line segment pointed on one end. (See Volume 0 for details.) Let C denote this bound vector.

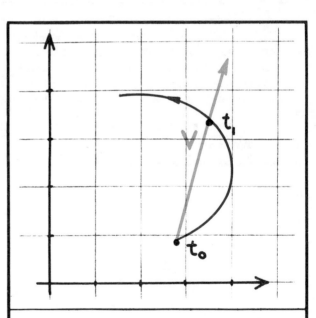

1.2.2. The *average velocity* of the change of state, C, is the vector starting at the point labeled t^0 on the curve, and directed along the vector of change of state, C, but divided by T, the time elapsed between t^0 and t^1. Let V denote this vector, $V=C/T$. It represents the average speed and direction of the change of state.

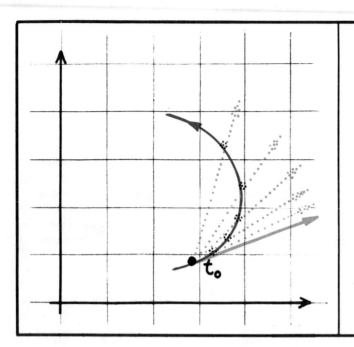

1.2.3. The *instantaneous velocity* of the trajectory at the time t_0 is the bound vector that V tends to, as the elapsed time, T, shrinks smaller and smaller. This limiting vector, denoted here by C,↑ is also known as the *tangent vector*. The construction of this velocity, or tangent vector, from the curve is called *differentiation* in Vector Calculus.

The modeling process begins with the choice of a particular state space in which to represent observations of the system. Prolonged observations lead to many trajectories within the state space. At any point on any of these curves, a velocity vector may be derived. This is the new dynamical concept of Newton and Leibniz. It is useful in describing an inherent tendency of the system to move with a habitual velocity, at particular points in the state space.

The prescription of a velocity vector at each point in the state space is called a *velocity vectorfield*.

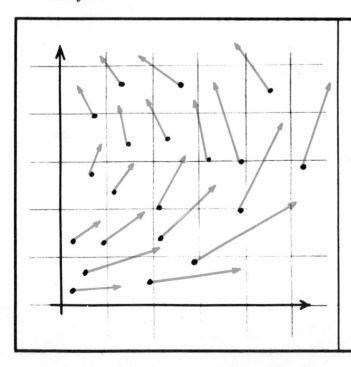

1.2.4. A *vectorfield* is a field of bound vectors, one defined at (and bound to) each and every point of the state space. Here, only a few of the vectors are drawn, to suggest the full field.

The state space, filled with trajectories, is called the *phase portrait* of the dynamical system. The velocity vectorfield has been derived from the phase portrait by *differentiation*.

We regard this vectorfield as the model for the system under study.
In fact, the phrase *dynamical system* will specifically denote this vectorfield.

In the practice of this modeling art, the choice of a vectorfield is a difficult and critical step. Extensive observations of the organism being modeled, over a long period of time, will usually reveal tendencies (to a dynamicist, at least) which can be represented as a dynamical system. The history of applied dynamics provides excellent examples of this process. Several of these are described in the next four chapters. The usefulness of this kind of model depends on the following fundamental hypotheses.

Hypothesis 1. The observation of the organism over time, represented as a trajectory in the state space, will have this property, at each of its points: its velocity vector is exactly the same as the vector specified by the dynamical system.

Henceforward, the word *trajectory* will always carry this assumption. That is, the trajectories of the phase portrait have the specified velocity vectors, and further, they will be assumed to represent the behavior of the system being modeled. Further, for technical reasons we also assume:

Hypothesis 2. The vectorfield of the model is smooth.

The word smooth, in this context, is most easily seen in the one-dimensional case. On a one-dimensional state space, a vectorfield is specified by a graph in the plane. In this context, the graph is *smooth* if it is continuous, and its derivative is continuous as well: no jumps, no sharp corners. More details are given in Volume 0 of this series.

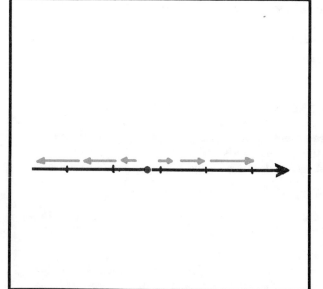

1.2.5. For example, here is a vector-field (green) on a one-dimensitonal state space (black). The vector at the rest point (red) is the "zero vector", its length is zero.

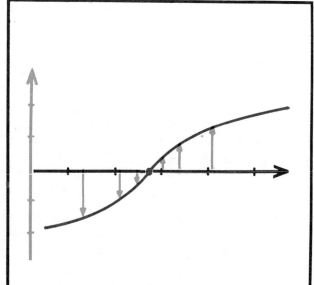

1.2.6. Stand up each green vector by means of a counterclockwise rotation by a right angle. The arrowheads (green) trace out a curve (red) which is the graph of a function. The vectorfield is completely described by this function.

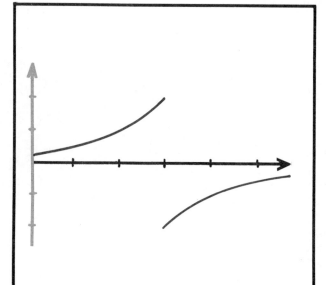

1.2.7. Another vectorfield is described by this function. This function is not continuous, so the vectorfield is not smooth.

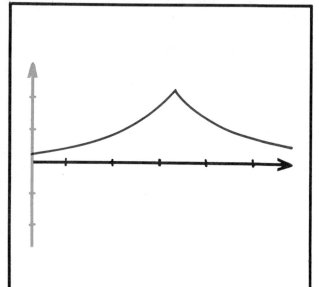

1.2.8. Yet another vectorfield is described by this function. This function is continuous, but has a sharp corner. This vectorfield is not smooth either.

We suppose now that a dynamical system has been chosen as a model for a system. Given this vectorfield, how can we deduce the trajectories, thus the phase portrait, and the behavior of the system?

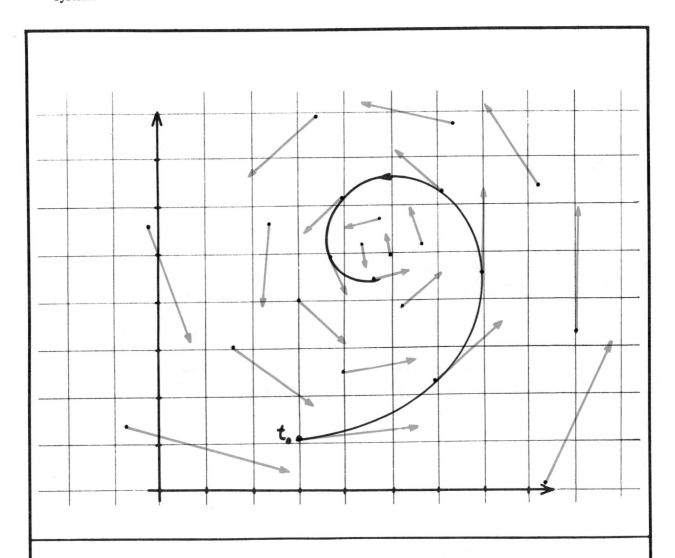

1.2.9. Given a state space and a dynamical system (smooth vectorfield), a curve in the state space is a *trajectory,* or *integral curve,* of the dynamical system if its velocity vector agrees with the vectorfield, at each point along the curve. This means the curve must evolve so as to be tangent to the vectorfield at each point, as shown here. The point on the trajectory corresponding to elapsed time zero, t=0, is the *initial state* of the trajectory.

Given a dynamical system (a smooth vectorfield on a manifold), how can we find its trajectories? Analysis, the mathematical theory which has evolved since Newton and Leibniz, has established that from each initial point, there is a single trajectory of the system. Finding it requires the construction called *integration* in Vector Calculus. Thus, trajectories are sometimes called *integral curves.*

A graphical construction which approximates the integration of a trajectory, or integral curve, was discovered by Euler.

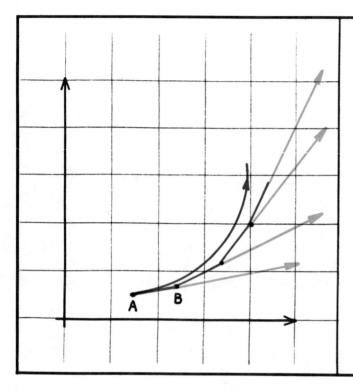

1.2.10. Euler's method approximates an integral curve by a polygon. Starting from the initial point, *A,* a straight line is drawn along the vector of the dynamical system attached to that point, *V(A).* The length of this straight line is a small proportion of the length of this vector, say 10%. At point *B,* at the far end of this line segment, the construction is repeated, using the vector, *V(B),* attached to this point by the dynamical system. This construction is repeated as many times as necessary, to draw the polygonal, approximate trajectory.

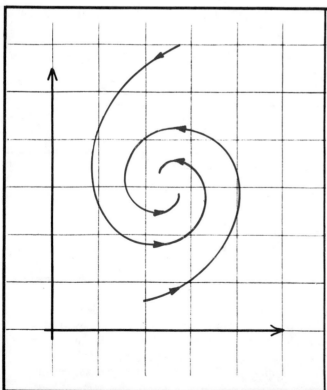

1.2.11. The state space is filled with trajectories, completely determined by the dynamical system. The display of the state space, decomposed into these curves, is the *phase portrait* of the system. Furthermore, the space of states may be imagined to flow, as a fluid, around in itself, with each point following one of these curves. This motion of the space within itself is called the *flow* of the dynamical system.

In the next chapter, simple examples will show that flat, Euclidean, spaces will not suffice for all of our geometric models. In some cases, curved spaces (that is, *manifolds)* will be necessary. In Global Analysis, the Calculus of Newton and Leibniz is generalized to the context of manifolds. This generalization provides the basic tools of mathematical dynamics. A description may be found, in pictorial representations, in Volume 0 of this series.

The trajectory and velocity vector concepts fit nicely into the context of manifolds. Here we illustrate these ideas on a two-dimensional manifold, which is simply a curved surface.

1.2.12. Here the instantaneous velocity vector is obtained as a limit of average velocity vectors, as in an earlier illustration. But in this case, the state space is *curved.* The velocity vector does not live in the curved surface. It sticks out into the ambient three-dimensional space. It is *tangent* to the surface.

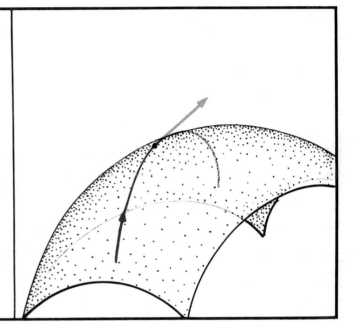

1.2.13. Now repeat this construction many times, with different curves lying in the surface, all passing through the same point. All the vectors lie in the same plane, tangent to the curved surface at a point. This plane is the *tangent space* of the space of states, at that point.

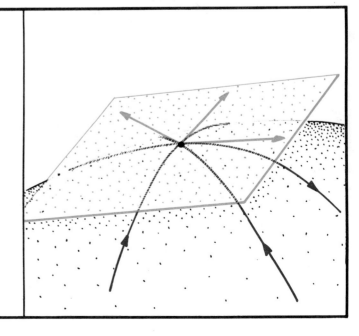

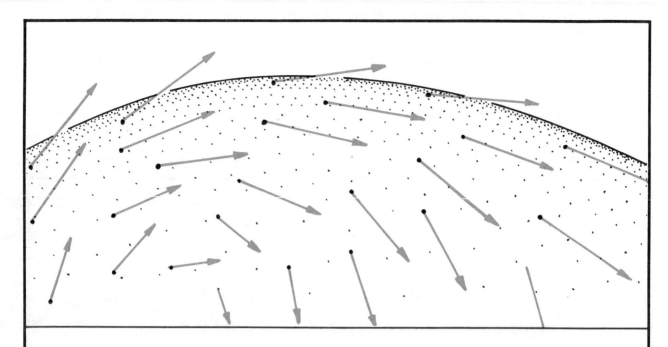

1.2.14. A vectorfield, in this context, means the assignment of a tangent vector to every point of the curved surface.

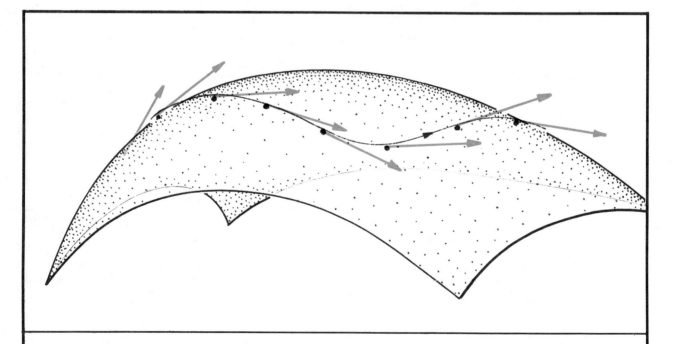

1.2.15. This is a trajectory, or integral curve, of a vectorfield (dynamical system) on a curved space of states. The tangent vectors at each point project off the surface, yet the integral curve stays within it.

This completes the introduction of the basic concept of a dynamical system, and the modeling process, which we may summarize as follows.

Suppose a dynamical model has been proposed for some experimental situation. This situation may be a laboratory device, an organism, a social group, or whatever. The model consists of a manifold and a vectorfield. The manifold is a geometrical model for the observed states of the experimental situation, and is called the *state space* of the model. The vectorfield is a model for the habitual tendencies of the situation to evolve from one state to another, and is called the *dynamic* of the model. Now mathematics can be brought out of the tool-box, and used to draw many trajectories of the dynamical system, creating its *phase portrait.* These basic concepts of dynamical systems theory have been illustrated in the preceding two sections.

Now you may ask: SO WHAT? Well, according to our *conventional interpretation,* the agreed rules of the game, these trajectories are supposed to describe the behavior of the system, as observed over an interval of time. Either they do this, with an accuracy sufficient to impress you, and be useful for predicting the behavior of the experimental situation, or they do not. In many examples of this art, called *applied dynamics,* they do. These models succeed remarkably well, and have been used by many satisfied customers over the years. Some of these examples are presented in the next four chapters.

But some obstinate reader's may still exclaim: SO WHAT? Well, dynamical systems theory has yet more to offer: PREDICTION FOREVER. Sophisticated techniques from the research frontier of pure mathematics have been employed to yield *qualitative predictions of the asymptotic behavior of the system in the long run, or even forever.* Although qualitative predictions are not as precise as quantitative ones, they are a whole lot better than no predictions at all. *And for most problems of applied dynamics, quantitative predictions are impossible.* So, the remaining sections of this chapter are devoted to the illustration of the concepts of *asymptotic behavior.*

1.3 SPECIAL TRAJECTORIES

The first step in the quest for qualitative predictions of asymptotic behavior is the examination of the phase portrait for special types of trajectories. Here we illustrate some of these special trajectories.

The simplest special trajectory is a point. Let's consider this in a one-dimensional context first.

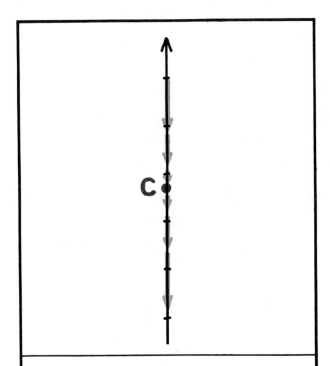

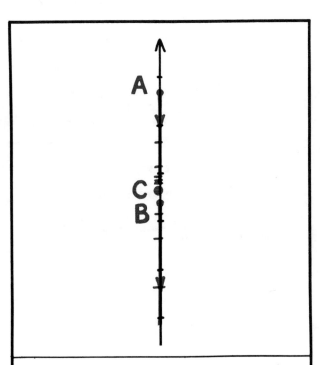

1.3.1. Here is a vectorfield on a one-dimensional state space. At one point in the state space, marked *C*, the associated velocity vector is the *zero vector*. This vector has length zero. The point, marked *C*, is called a *critical point*, or an *equilibrium point*, of the vectorfield. Because we assumed at the start that the vectorfield is *continuous*, the velocity vectors attached to points near the critical point are very short. (This is explained in Volume 0.)

1.3.2. This is the phase portrait of the dynamical system to the left. Three trajectories are shown, starting at the points, *A, B,* and *C.* Tick marks along the trajectories indicate the positions at successive seconds. Note that they are closer together near the critical point. One trajectory is piled up on the critical point. The velocity of this trajectory is zero, at all times. It does not move. It is called a *constant trajectory.*

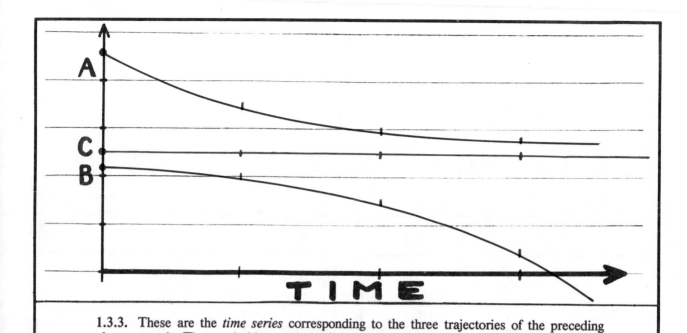

1.3.3. These are the *time series* corresponding to the three trajectories of the preceding phase portrait. The graph (time series) of the trajectory of *C* is a horizontal line, a constant function of time. This represents the constant trajectory of the critical point.

Now let's look at the same idea in a two-dimensional context.

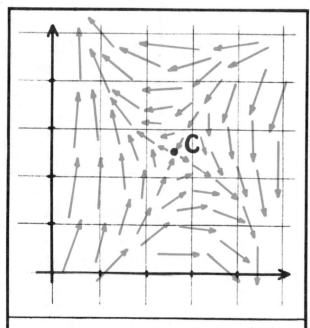

1.3.4. This is a garden-variety vectorfield in the plane. The zero vector appears once, as the velocity vector of the critical point, *C*. Nearby, the vectors are short.

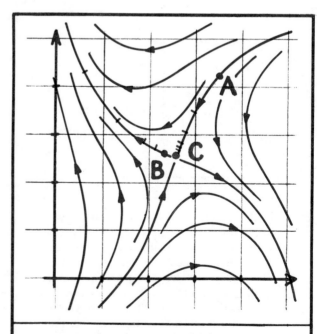

1.3.5. This is the phase portrait of the vectorfield to the left. The trajectory of the critical point is again piled up on the critical point. It is a constant trajectory.

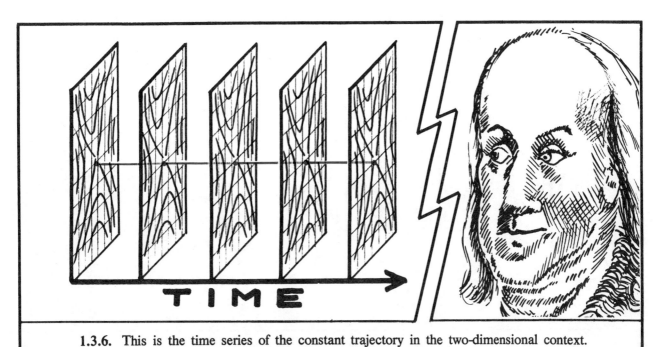

1.3.6. This is the time series of the constant trajectory in the two-dimensional context. Once again, it is a constant function of the time parameter.

In two dimensions or more, other types of special trajectories frequently occur. Here is a very important one, the *cycle*.

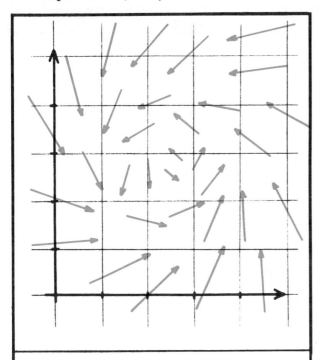

1.3.7. This planar vectorfield has an *eddy*. It seems as if the flow must somehow circle about a point.

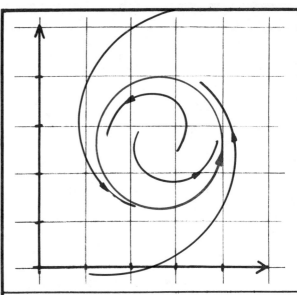

1.3.8. This is the phase portrait of the preceding dynamical system. Indeed, here we find a trajectory which is wrapped around and around the same curve. This is called a *closed trajectory*. It is also known as a *closed orbit*, *periodic trajectory*, *cycle*, or *oscillation*.

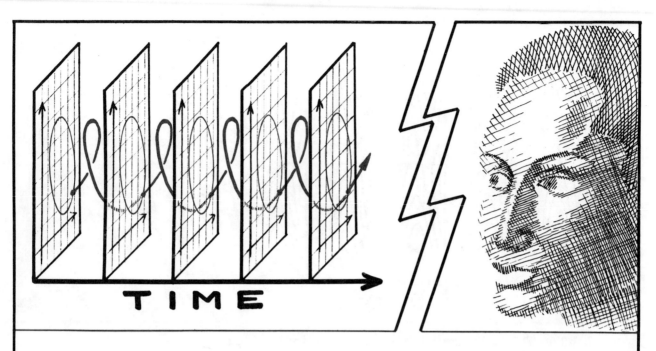

1.3.9. This is the time series representation of a closed trajectory. The graph wraps around a horizontal cylinder. The same interval of time is required to complete each wrap. This time interval is called the *period* of the closed trajectory.

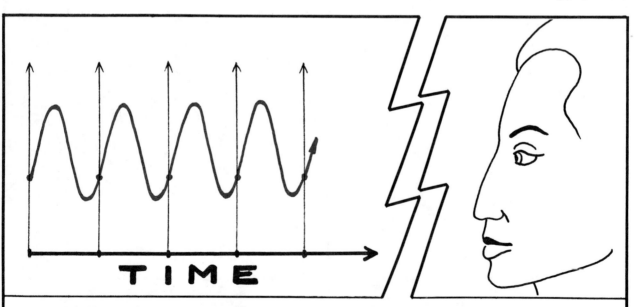

1.3.10. If a single parameter (of the two coordinates in the plane) is chosen, and the other data are forgotten, the time series of the chosen data may be plotted in the plane. The result, the *time series of the preferred parameter,* is a *periodic function.* This means that in every vertical strip corresponding to one period (or wrap, or cycle) of the closed trajectory, the graph exactly repeats itself.

These special types of trajectories, constant and periodic trajectories, also occur in phase portraits of dimension three or more.

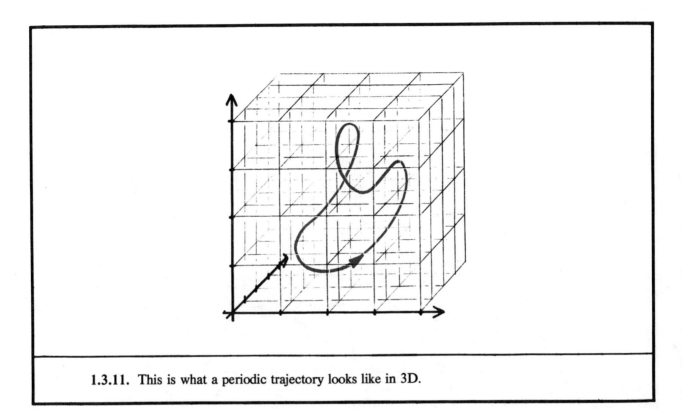

1.3.11. This is what a periodic trajectory looks like in 3D.

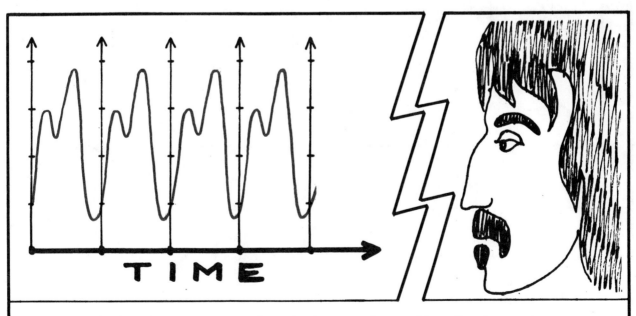

1.3.12. Choosing one parameter from the three coordinates, this preferred parameter may be recorded along the periodic trajectory. The time series of these data is again a periodic function.

But in higher dimensional phase portraits, other special trajectories may be found.

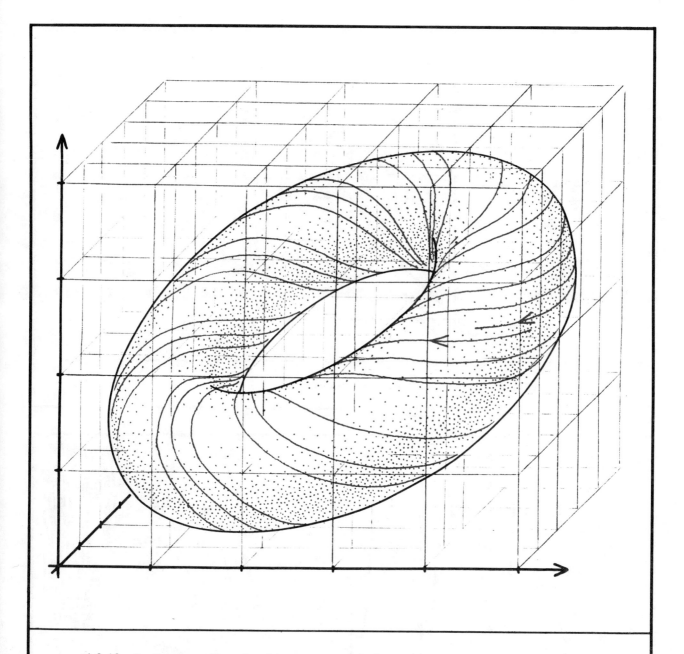

1.3.13. In the three-dimensional state space, imagine a torus (doughnut-like surface, see Volume 0 for details) with an infinitely long coil of wire wrapped endlessly around it, but never crossing itself or piling up. This can occur as a trajectory of a dynamical system, as we shall see in Section 4.4. It is called a *solenoidal* or *almost periodic trajectory*. An application is discussed in Chapter 5.

This finishes our list of special types of trajectory. Their significance in applications depends on the fact they they occur as limit sets, as we describe in the next section.

1.4   ASYMPTOTIC APPROACH TO LIMIT SETS

The second step in the dynamical systems quest for qualitative predictions of asymptotic behavior is the examination of the phase portrait for *asymptotic limit sets*. Let's see what this means.

We reconsider critical points in one dimension first, to start with the simplest case.

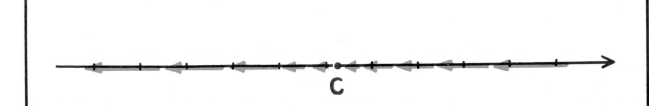

1.4.1. Here is the vectorfield on a one-dimensional state space. Recall that the point marked C is a critical point of the vectorfield. Because the vectorfield is smooth, the velocity vectors attached to points near the critical point are very short.

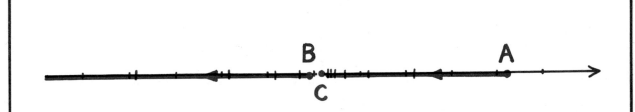

1.4.2. This is the phase portrait of the dynamical system above. Two trajectories are shown, starting at the points, A, and C. Tick marks along the trajectories indicate the positions at successive seconds. Note, again, that they are closer together near the critical point. The trajectory of C is a constant trajectory, piled up on the critical point. As time marches on, the trajectory of A gets ever closer to the point C. As it gets closer, it slows down. It gets closer and slower indefinitely, and approaches the critical point *asymptotically*. That is, it takes forever to reach C. We say that C is the *limit point* of the trajectory through A.

This trajectory approaches its limit point asymptotically.

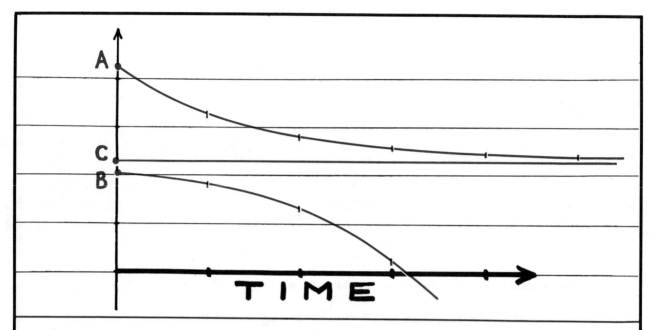

1.4.3. These are the time series corresponding to the three trajectories of the preceding phase portrait. The graph (time series) of the trajectory of *C* is a horizontal line, a constant function of time. This represents the constant trajectory of the critical point. The graph of the trajectory of *A* is descending to the right toward the horizontal line. It approaches this line *asymptotically,* as time increases to the right. The horizontal line is the *asymptote* of the graph of the trajectory of point *A*.

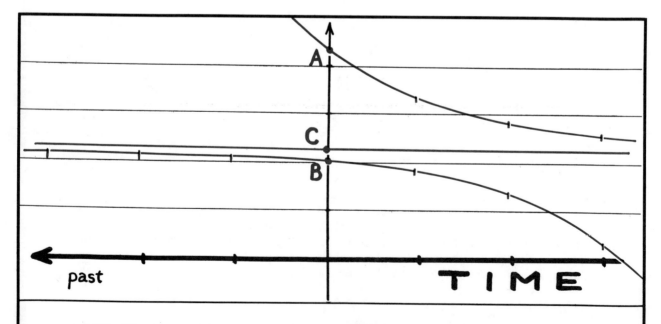

1.4.4. The graph of the trajectory of *B* similarly approaches asymptotically to the horizontal line, but going *backwards in time*.

Now let's look at these ideas in a two-dimensional context.

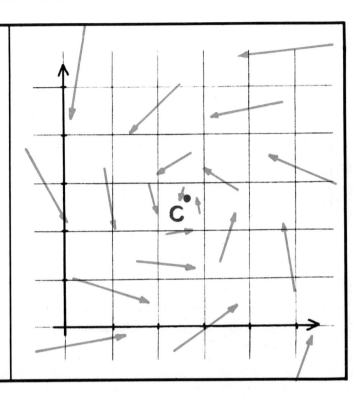

1.4.5. This is a garden-variety vectorfield in the plane. As in the planar example in the preceding section, the zero vector appears once, at the point *C*. Nearby, the vectors are short. But this vectorfield is different. The vectors spiral around the critical point. It is a critical point of *spiral type*.

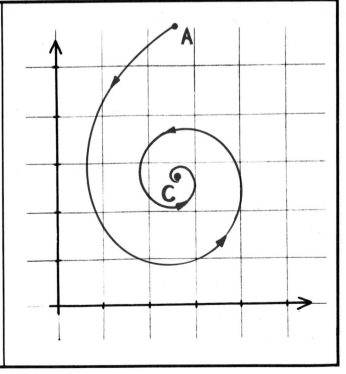

1.4.6. This is the phase portrait of the vectorfield to the left. The trajectory of the critical point is again piled up on the critical point. It is a constant trajectory. The other nearby trajectories, such as the one through the point marked *A*, spiral around the critical point, getting closer and closer. They approach this point *asymptotically*, slowing down as they close in. We say that the critical point, *C*, is a *limit point* of the trajectory through the point *A*.

This trajectory approaches its limit point asymptotically.

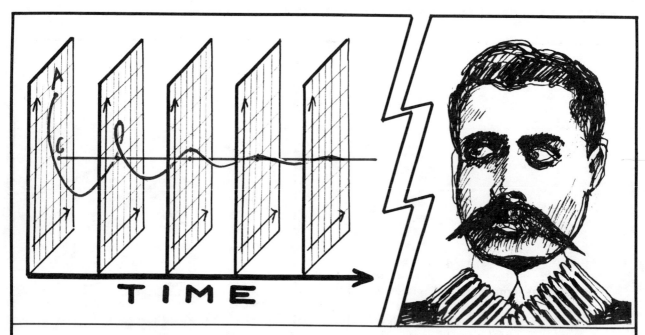

1.4.7. Here are the time series of two trajectories of this phase portrait. The time series of the constant trajectory, piled up on the critical point, C, is the graph of a constant (vector-valued) function. This graph is a horizontal, straight line. The time series of the nearby point, A, spirals around this straight line, approaching closer and closer as time moves to the right.

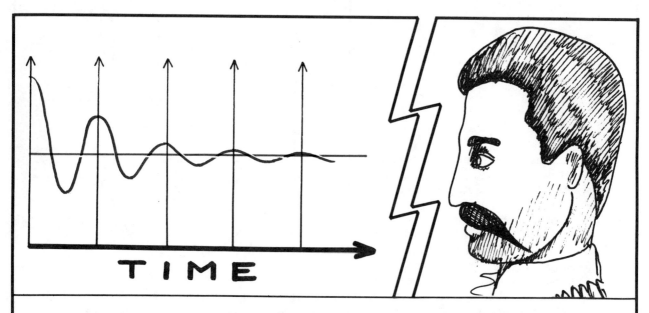

1.4.8. Choosing one of the two coordinates of the plane as a preferred parameter, the time series of this parameter along the two trajectories looks like this. The time series of the trajectory through A (wavy curve) approaches asymptotically toward the time series of the constant trajectory (horizontal, straight line) as time increases to the right.

In two dimensions or more, other types of special trajectories frequently occur. One of these, as we have seen in the previous section, is the *cycle*. A cycle may be the asymptotic limit set for a trajectory.

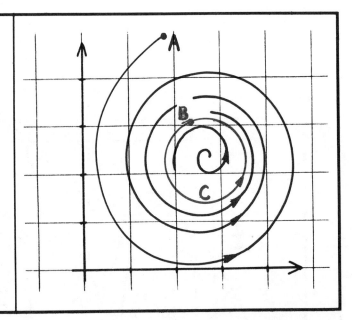

1.4.9. This is the phase portrait of a planar vectorfield with a cycle, marked *C*. A point on the cycle is marked *B*. The trajectory through *B* is a closed trajectory, winding around and around this cycle. Another trajectory is shown, through the point marked *A*. This trajectory spirals around the cycle, getting closer and closer as time goes on. We say that *C* is a *limit cycle*. It is the *limit set* for the trajectory through the point *A*.

1.4.10. This is the time series representation of the two trajectories. A horizontal cylinder is shown, extending to the right from the cycle, *C*. The time series for the closed trajectory, *B*, wraps around this horizontal cylinder. The time series for the spiraling trajectory, *A*, is wound loosely around the cylinder, and gets tighter and tighter as time increases, to the right.

Limit points and limit cycles also occur in phase portraits of higher dimensions. Further, in dimensions greater than two, other limit sets may turn up. For example, a torus can occur as a limit set, in a three-dimensional system. The solenoid, described in the preceding section, is a case in point.

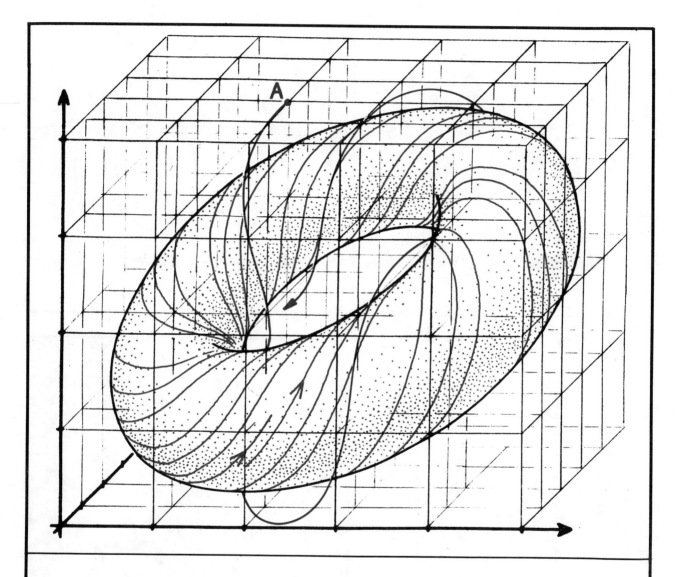

1.4.11. Here, the trajectory through the point marked *A* is wound, like a loose solenoid, around the torus. As time goes on, it winds around tighter and tighter. It approaches its limit set, the torus, asymptotically.

There are many more limit sets. Some of the more exotic ones will be shown in Volume 2. But, we have not yet made our case for the importance of the geometric theory of dynamical systems. To explain what we mean by *prediction forever,* a further concept is needed: that of an *attractor.* This is an outstanding type of limit set. It represents the behavior of a system in *dynamical equilibrium, after transients die away.* So let's go on.

1.5 ATTRACTORS, BASINS, AND SEPARATRICES

If an organism is dropped into a prepared environment, or an experimental device is prepared in an initial state and then turned on, we expect to see a brief settling-in period, before it settles down to an observable behavior. The erratic behavior during the initial settling-in period is called the *start-up transient*. The settled-in, eventual observable behavior is the *equilibrium state* of the experiment. *Warning: Equilibrium, as used here, does not imply a static equilibrium, nor a steady state.*

In a dynamical system modeling this experimental situation, a trajectory will model the start-up transient, while its limit set models the equilibrium state which follows. The asymptotic approach of the trajectory to its limit set models the dying away of the transient, as the system settles to its dynamic equilibrium.

For probability reasons to be explained shortly, the only equilibrium states which may be observed experimentally are those modeled by the limit sets which receive most of the trajectories. These are called *attractors*.

Here is the attractor concept, illustrated in two dimensions. The same ideas apply in all dimensions. First, we consider limit points, the simplest limit sets.

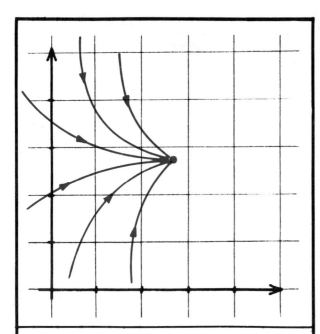

1.5.1. Suppose a dynamical system in the plane has a critical point. And let's suppose further that this critical point is the limit set of some trajectories in the phase portrait.

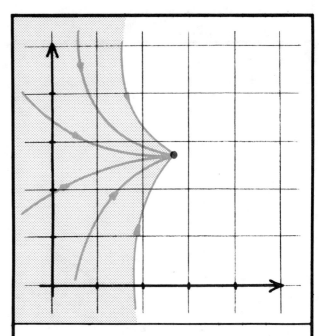

1.5.2. Now, find every single trajectory which approaches this limit point asymptotically, and color it green. The green portion of the plane is the *inset* of the limit set (that is, the critical point).

The inset of a limit set represents, in a dynamical model, all the initial states which end up in the same equilibrium state, after the start-up transient dies away.

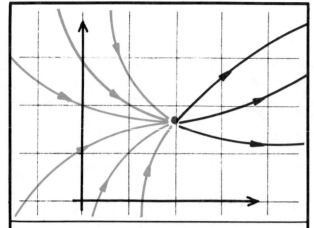

1.5.3. Other trajectories depart from the limit point. That is, *if the direction of time were reversed, these trajectories would approach asymptotically to this limit point.* Restoring the direction of time to normal again, we say these departing trajectories have the critical point as their *alpha-limit set.*

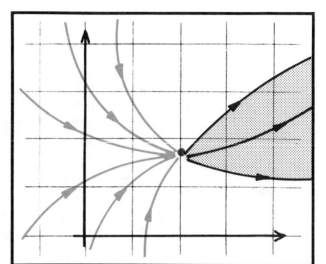

1.5.4. Now find every trajectory which has this critical point as its alpha-limit set, and color it blue. The blue portion of the plane is the *outset* of the limit set (that is, the critical point.)

Sometimes we say *omega-limit,* in place of just plain limit. Then omega-limit set refers to the future asymptotic behavior, while alpha-limit set refers to the past. The trajectory goes from alpha-limit to omega-limit.

Warning: Some trajectories neither arrive nor depart at a critical point, although they may pass closely by!

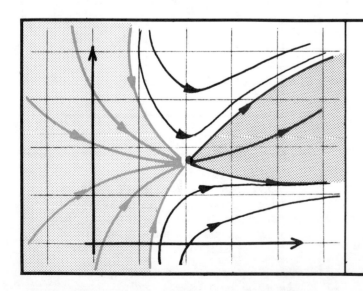

1.5.5. Here we see the same dynamical system in the plane. Both the inset and the outset are colored-in. The black trajectories start near the inset, and go towards the limit set for a while. When they get near the limit point, they feel the influence of the outset, and turn aside. Following the outset, they disappear into the distance.

When a trajectory flies by, it is on its way somewhere. Either it has an omega-limit set elsewhere in the phase portrait, or it departs from the state space, never to be seen again. Similarly, it came from somewhere else. Thus, every trajectory may belong simultaneously to the outset of one limit set (its alpha-limit set) and to the inset of another (its omega-limit set.) In two dimensions, cycles may be limit sets. Limit cycles have insets and outsets, too. (So do the other limit sets, in higher dimensions.)

What if all nearby trajectories are arriving?

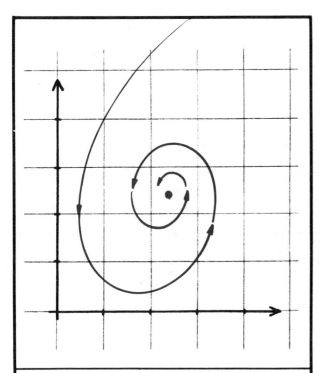

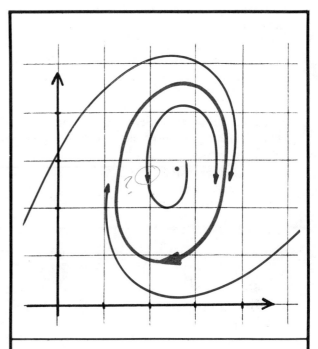

1.5.6. Here is a limit point in two dimensions. Every nearby trajectory is arriving. The Inset contains an open disk around the limit point. Every initial point in this disk is captured by the limit point, when its transient dies away. If an initial point is chosen at random from all the points in the state space, the probability of it asymptotically approaching this limit point, instead of some other, is positive. (In fact, it is 100% in this particular example.) This limit point is an *attractor*. As it is a point, and represents static equilibrium, it is also called a *static attractor*.

1.5.7. This is the phase portrait of a different dynamical system in the plane. It has a limit cycle. Again, the inset of this limit set is as large as possible. Except for the solitary red point in the center, which is a critical point, every single initial state evolves to the same limit set. The inset includes an open annulus (ring) around the limit cycle. The probability that an initial state, chosen at random from among all the initial states in the state space, will end up at this limit cycle is positive. (In fact, it is 100% in this particular case.) This limit cycle is an attractor. As it is a cycle, and represents a periodic equilibrium, it is also called a *periodic attractor*.

An *attractor* is a limit set with an *open inset*. **That is, there is an open neighborhood of the limit set within its inset. (Refer to Volume 0 if** *open set* **is unfamiliar.)**

Of all limit sets, which represent possible dynamical equilibria of the system, the attractors are the most prominent, experimentally. This is because the probability of an initial state of the experiment to evolve asymptotically to a limit set is proportional to the volume of its inset. We will say that a limit set is *probable* if the volume of its inset is a positive number, instead of zero. Open sets have positive volume, although not every set of positive volume is an open set. Attractors have open insets, so they are probable. They are experimentally discoverable. Other limit sets may be probable, without being attractors in the strict sense of the preceding paragraph. They are called *vague attractors*. This is synonymous with *probable limit set*. These are also experimentally discoverable. Limit sets which have *thin insets* (that is, that have probability zero) are *non-attractors*. They are experimentally insignificant. These are called *exceptional limit sets,* or synonymously, *improbable limit sets*[2].

The inset of an attractor is called its *basin*. In a typical phase portrait, there will be more than one attractor. The phase portrait will be divided into their different basins. The dividing boundaries (or regions) are called *separatrices*. In fact, any point which is not in the basin of an attractor belongs to a separatrix, by definition.

Here are some examples of attractors, basins, and separatrices, in two dimensions. The same concepts apply in three or more dimensions, but are harder to visualize.

COLOR CONVENTION:
    Attractors = red
    Basins = blue
    Separatrices = green

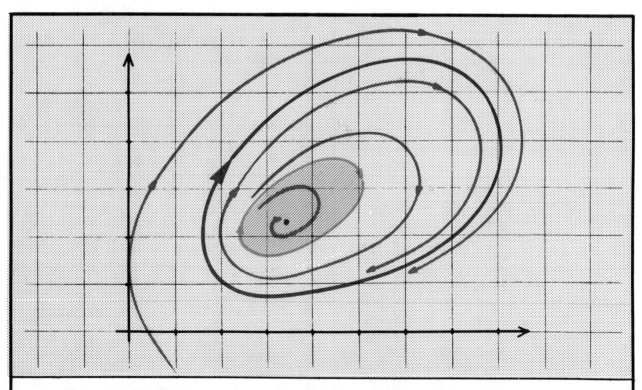

1.5.8. In this example, there are two attractors: a point and a cycle. A third limit set, a cycle, comprises the separatrix.

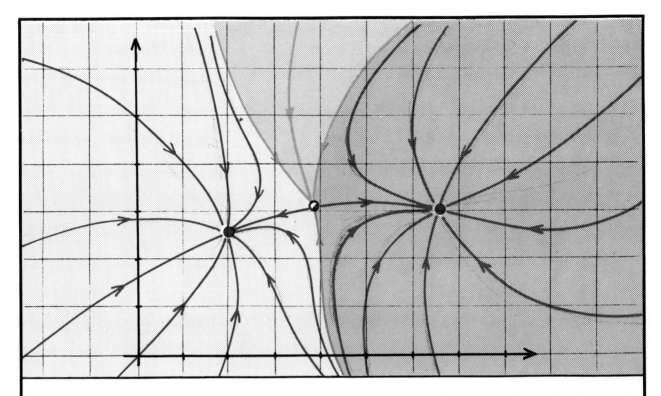

1.5.9. Here, also, there are two attractors: both points. A third limit set, also a point, is not an attractor. It is an exceptional limit set. The inset of this exceptional limit set comprises the separatrix.

This reveals the pattern of the general case: the separatrix consists of all points not in a basin. Every point tends to a limit set. If its limit set is an attractor, it belongs to a basin. So, if it belongs to the separatrix (and therefore not to a basin), it must tend to a non-attractor. Thus: *the separatrix consists of insets of exceptional limit sets.*

The preceding examples are artificial, made up to illustrate the concepts. But, we are overdue for some more meaningful examples. So, at this point, let's turn to gradient systems—a rich source of simple examples, based on a geometrical construction.

1.6   GRADIENT SYSTEMS

The *gradient* operation of vector calculus provides dynamical systems (vectorfields) of an especially simple type called a *gradient system*. In these, there is an auxiliary function, called the *potential function*. The velocity vectorfield is simply the *gradient vectorfield* of this potential function.

This section develops a typical example of gradient dynamics in the plane.

1.6.1.   The state space, in this example, is the plane. The potential function is a function from the state space to the real number line. To each point in the state space, it assigns a real number. This number is the *potential* of the corresponding state. In an application of this scheme, this potential would presumably be observable, or deducible from observations.

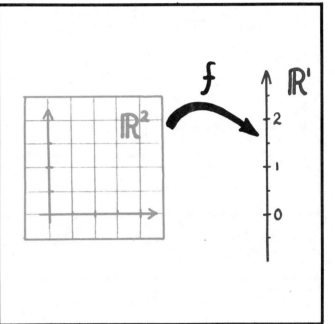

1.6.2.   Represent this function as a graph in three-dimensional space. The state space is the horizontal coordinate plane. From each point in this plane, move vertically a distance equal to the potential of that point—up if the potential is a positive real number, down if it is negative.

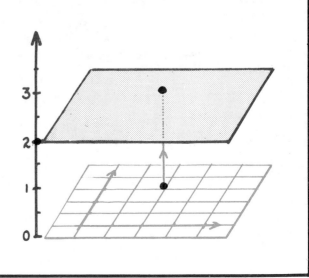

The graph of a potential function on a planar state space is a surface in three-dimensional space, called the *potential surface*. We may think of this as a landscape.

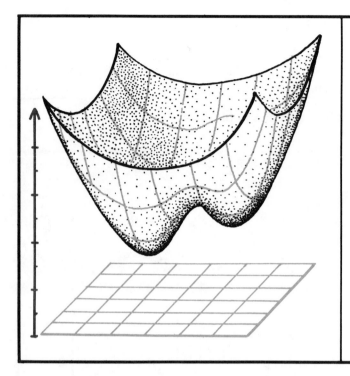

1.6.3. For the sake of definiteness, let's choose a particular potential function, and visualize it as a potential surface. This one, for example, has two valleys, with a saddle ridge in between.

An alternate representation of a function is its *contour map*. Let us represent our exemplary potential this way.

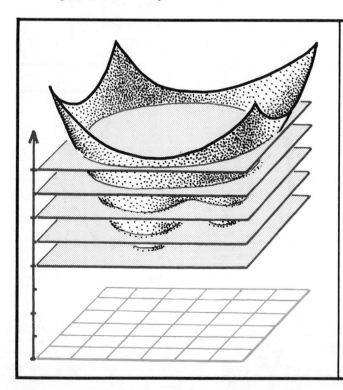

1.6.4. At regular intervals along the vertical axis, draw horizontal, cutting planes. Mark the potential surface with a red curve, where each cutting plane cuts through the surface. These are called the *level curves* of the surface.

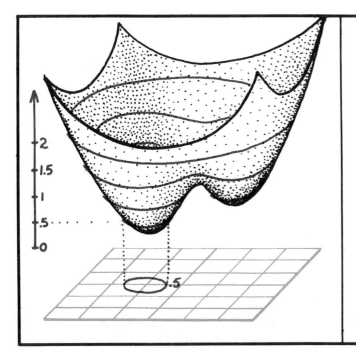

1.6.5. Next, each level curve is projected onto the horizontal coordinate plane. This curve in the state space, also called a *contour curve,* contains every state with the same value of the potential function. Over this curve, the potential surface has a constant height. Label this contour curve with its common value of the potential.

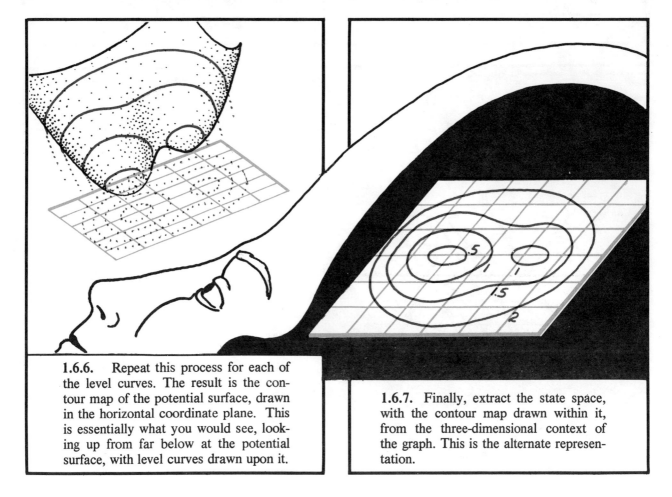

1.6.6. Repeat this process for each of the level curves. The result is the contour map of the potential surface, drawn in the horizontal coordinate plane. This is essentially what you would see, looking up from far below at the potential surface, with level curves drawn upon it.

1.6.7. Finally, extract the state space, with the contour map drawn within it, from the three-dimensional context of the graph. This is the alternate representation.

The gradient dynamical system for this particular potential function is derived as follows.

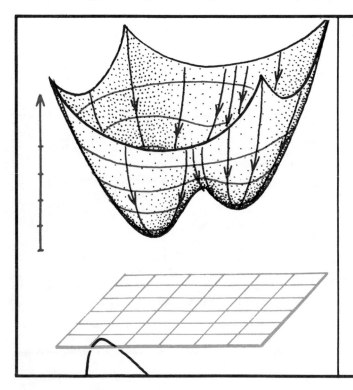

1.6.8.   Sprinkle the potential surface with a fine mist of blue ink. Droplets will run down the fall-lines, that is, the routes of steepest descent. Suppose that the speed of a droplet is exactly the steepness of the slope.

1.6.9.   Viewed from far below, the blue droplets appear to move over the contour map, at right angles to the contours.

The blue curves, perpendicular to the contours, together with the parameter of time along each, comprise the phase portrait of the gradient dynamical system.

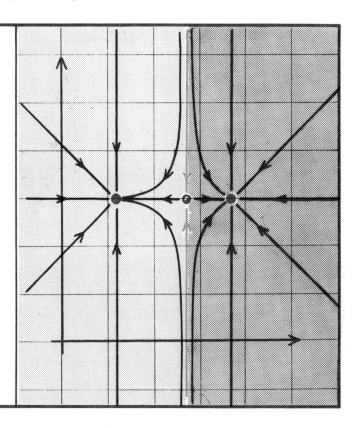

1.6.10. In this example, the gradient phase portrait has two basins, with a point attractor in each. Between them is a limit point of saddle type, or *saddle point,* corresponding to the saddle on the ridge between the two valleys in the landscape. The inset of the saddle point consists of the two green trajectories. This inset is also the separatrix, dividing the state space into the two basins.

Gradient systems, generally, are much like this example. Their limit sets are generally equilibrium points. A limit cycle is impossible in a gradient system, as you cannot go steadily downhill and still return to your starting point (except in an Escher print). Although gradient systems are useful in some elementary physics problems, their usefulness in general applications is severely limited by the lack of limit cycles. The next chapter will show why limit cycles are so important.

2. CLASSICAL APPLICATIONS: Limit points in 2D from Newton to Rayleigh

The early work in applied dynamics, before Rayleigh and Poincaré, was devoted primarily to the motions of the planets around our sun. This application, called *celestial mechanics,* comprises an enormous and important field. It was the main subject of Poincaré's research, and is still very active. Under the name *conservative mechanics,* it has been enlarged to include all non-dissipative, that is, frictionless, mechanical systems.

General dynamical systems in nature — whether physical, chemical, biological, or social ¬ are not conservative, but *dissipative.* The theory and applications of these dynamical systems have been elaborated mostly in this century, after the lead of Euler, Rayleigh, Poincaré and Liapounov.

The classical examples of dissipative dynamical systems — primarily those included in Rayleigh's text of 1877 — are described in this chapter. Some of the analytical descriptions, as systems of ordinary differential equations of first order written classically, are included here without discussion. Explanations of these expressions, not essential for understanding the rest of this text, may be found in the appendix.

2.1 PENDULA

The pendulum may be the most classical example of the dynamical modeling process. It has a two dimensional state space, and a dynamical system established by Newton.

This model assumes that the rod is very light, but rigid. The hinge at the top is perfectly frictionless. The weight at the lower end is heavy, but very small. It moves in a vacuum. The force of gravity always pulls it straight down.

These idealizations describe the modeling assumptions in this example, called the *simple pendulum.*

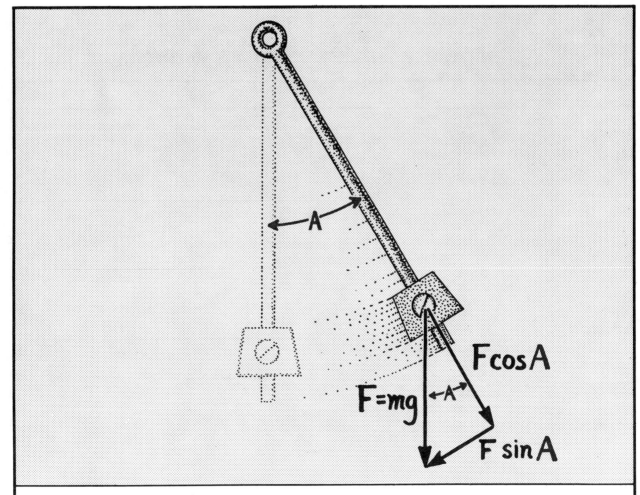

2.1.1. If *A* denotes the angle of elevation of the pendulum, and *F* the force of gravity, then *FcosA* is the pull along the rod, and *FsinA* the force turning it, as shown here.

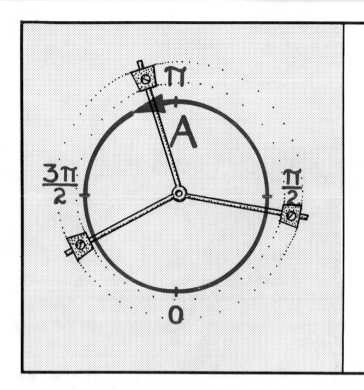

2.1.2. The angle of elevation, *A,* parametrizes a circle. That is, values of *A* can be any real number, but *A=0* and *A=2π* denote the same angle. The angle, *A,* represents a point of the circle. It is called an *angular variable.*

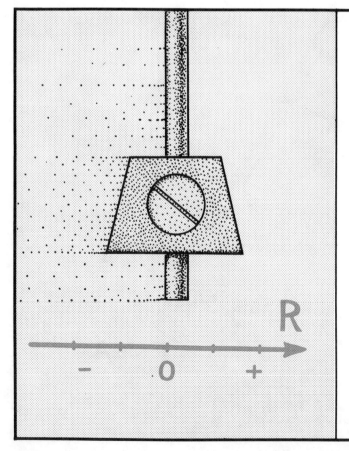

2.1.3. Let *R* denote the rate of rotation of the rod at a given moment. This rate is also observable, by radar for example. In Newton's model, this parameter is included, along with the angle, *A,* as a descriptor of the state of the pendulum. The rate of rotation, *R,* may have any real number as its value. It represents a point of the real number line.

2.1.4. The two parameters, *A* and *R*, together locate a point on a circular cylinder. This is the state space of Newton's model. The vertical circle in the center of this cylinder denotes the states of zero angular velocity, *R=0*. The straight line from front to back, at the bottom of the cylinder, is the axis of zero inclination, *A=0*, where the pendulum is lowest.

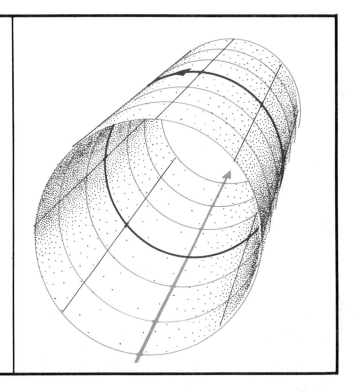

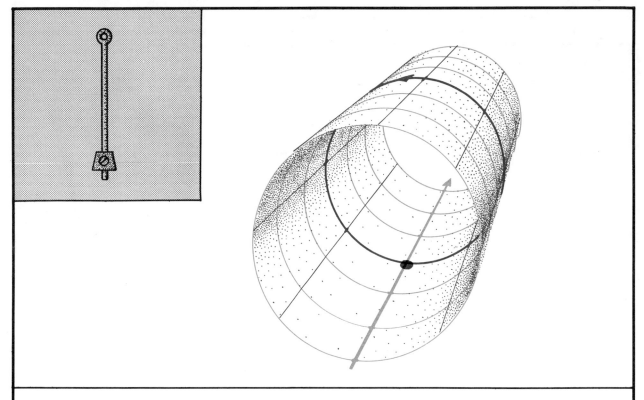

2.1.5. At the *origin* defined by *(A,R)=(0,0)*, the pendulum is at rest at its lowest position.

Moving the pendulum a little to the left and then letting it go with no shove causes it to swing back and forth indefinitely. Remember, there is no friction in the hinge and no air in the way. The representation of this motion as a trajectory in Newton's model is shown here in four steps.

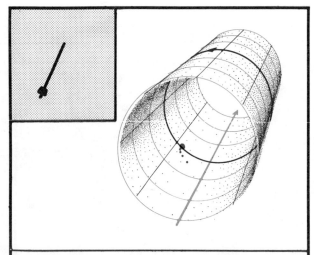

2.1.6. Step 1. Immediately after the pendulum is released, the representative point is on the circle of *R=0* to the left of the origin, moving away from us as the rate *R* increases.

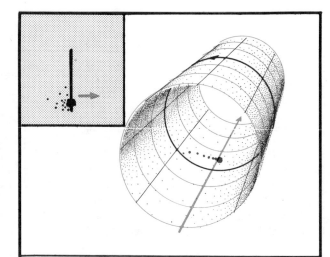

2.1.7. Step 2. It also moves to the right as the inclination increases. Here it has just reached the axis, A=0, as the pendulum goes by its bottom point.

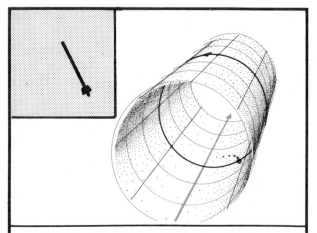

2.1.8. Step 3. It continues to move to the right, moving towards us rather than away, as *R* decreases. It reaches the circle of *R=0,* when the pendulum attains its maximum swing to the right, and turns to fall again toward its bottom, *A=0.*

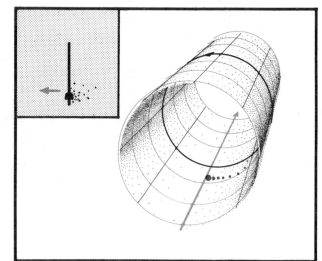

2.1.9. Step 4. It approaches us and moves to the left, as the pendulum falls. It crosses the axis, A=0, as the pendulum swings through bottom.

Then, the cycle begins again. The full trajectory in the state space, corresponding to this oscillating motion of the pendulum, is a cycle, or closed loop.

The next sequence shows the trajectory in Newton's cylindrical model, representing the motion of the pendulum, dislodged from a precarious balance at the top.

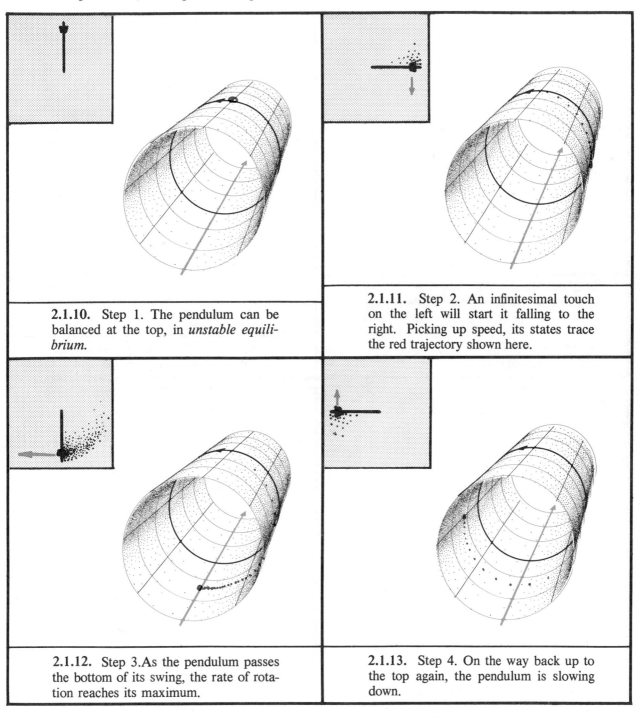

2.1.10. Step 1. The pendulum can be balanced at the top, in *unstable equilibrium.*

2.1.11. Step 2. An infinitesimal touch on the left will start it falling to the right. Picking up speed, its states trace the red trajectory shown here.

2.1.12. Step 3. As the pendulum passes the bottom of its swing, the rate of rotation reaches its maximum.

2.1.13. Step 4. On the way back up to the top again, the pendulum is slowing down.

As this trajectory approaches the position of balance at the top (its omega-limit point, an unstable equilibrium) it moves slower and slower. The pendulum actually balances at the top again, at the end of the motion, but this motion takes forever.
Warning. This trajectory is not a cycle, because the point at the top, the omega-limit point, does not belong to the trajectory.

The next sequence is a slight modification of the preceding–the pendulum is balanced at the top, then shoved hard to the right.

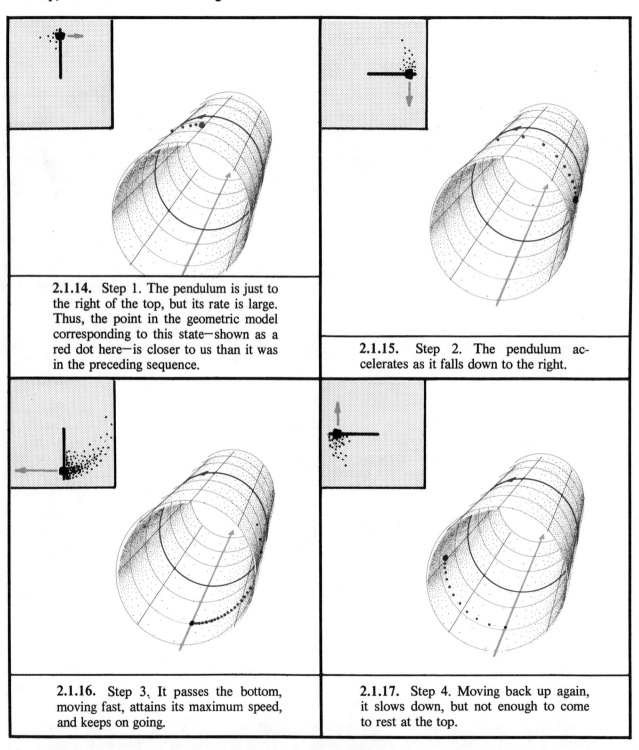

2.1.14. Step 1. The pendulum is just to the right of the top, but its rate is large. Thus, the point in the geometric model corresponding to this state—shown as a red dot here—is closer to us than it was in the preceding sequence.

2.1.15. Step 2. The pendulum accelerates as it falls down to the right.

2.1.16. Step 3. It passes the bottom, moving fast, attains its maximum speed, and keeps on going.

2.1.17. Step 4. Moving back up again, it slows down, but not enough to come to rest at the top.

In this motion, the pendulum rotates clockwise indefinitely. The corresponding (red) trajectory in the cylindrical state space closes at the top. It is a cycle. But unlike the slow oscillation described above, this fast cycle goes around the cylinder.

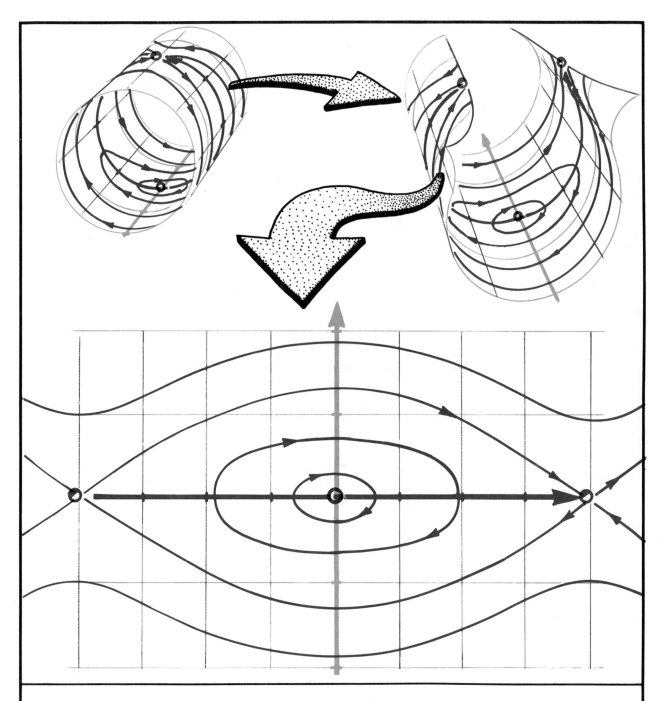

2.1.18. Performing many such experiments with a pendulum, or emulating one with an analog or digital computer, would reveal the phase portrait of Newton's model. This cylinder full of trajectories is easier to see if it is cut down the top, and flattened, as shown here. Notice that there are two equilibrium points. One at the top is a *saddle,* a type we have already seen in the gradient dynamical system in the preceding chapter. The other, at the origin, is another type called a *center,* or *vortex point.* This type will recur throughout the rest of this book. The critical point in the center is *not a limit point* of the nearby trajectories.

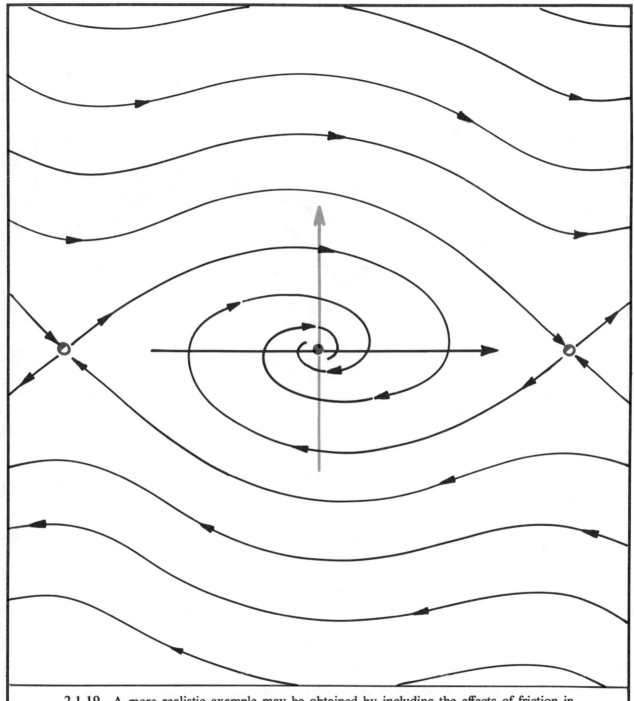

2.1.19. A more realistic example may be obtained by including the effects of friction in the hinge, and air resistance. Here is the phase portrait which results. Notice that it is very similar to the preceding portrait, but the equilibrium point at the origin is no longer a *center*. It has become an *attractor*. This is because any nearby trajectory, representing a slow motion of the pendulum near the bottom, will die away because of friction, and the pendulum will *come to rest*. This spiraling type of point attractor is sometimes called a *focal point*.

An instructive variation of this example may be produced by adding small forces which break the symmetry.

2.1.20. Suppose the pendulum is magnetic, and two magnets are added near the bottom of the arc. The one on the left is stronger.

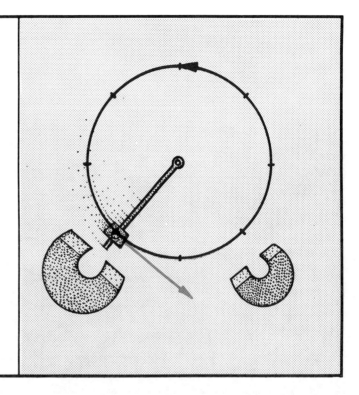

2.1.21. As the pendulum bob could be stopped near either magnet, and held slightly aloft by its attraction, the dynamical system modeling this device will have two basins, with a point attractor in each.

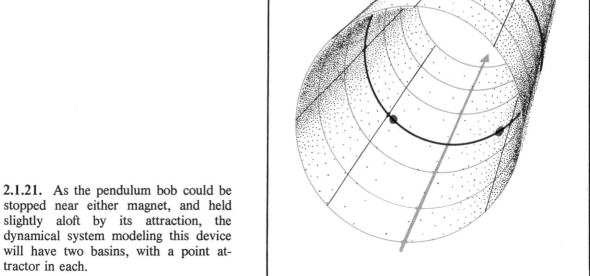

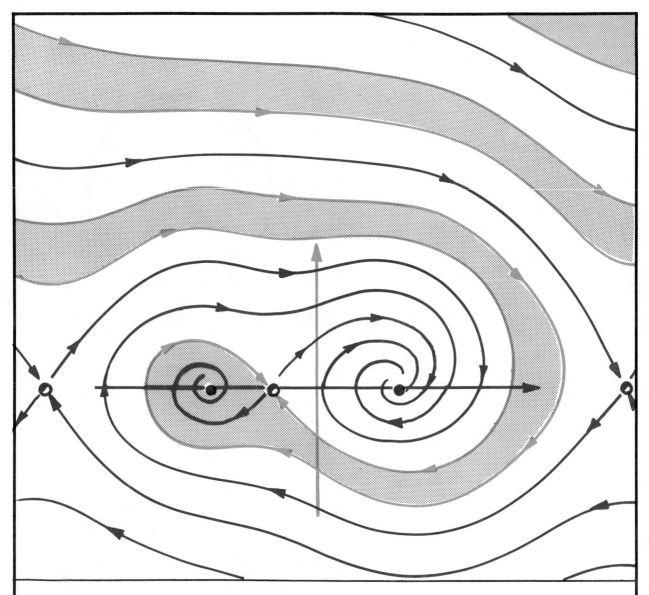

2.1.22. Here is the phase portrait of the magnetic pendulum, unrolled. The basin of the rest point near the smaller magnet, shaded, is smaller. Thus, the probability of an initial state (angular position plus angular velocity) evolving asymptotically to the rest point adjacent to the smaller magnet is less than 50%.

   Note. In this particular case, the shaded basin extends upward only. Thus, if the bob is swinging rapidly counterclockwise, it cannot come to rest at the smaller magnet.

This is not solely an armchair experiment. It has actually been studied[1].

This is the most classical application. Although pendulum theory may not turn everyone on, it has certainly been fundamental to the growth of dynamics. The writings of Lord Rayleigh are held together with this common thread, and he saw pendula everywhere. The spirit of his work lives on in many dynamics laboratories, even today.

2.2   BUCKLING COLUMNS

The deformations of elastic, solid material provided the context for some of the early applications of dynamics. The buckling of a vertical, elastic column – under a weight balanced on its top – was studied by Euler. A particularly simple analysis of this experimental situation has been made by Stoker[1]. His dynamical model, illustrated in this section, is closely related to the model for the simple pendulum, shown in the preceding section.

2.2.1. An *elastic column* – a thin metal bar balanced on end – is slightly compressed by a light weight on top. If the center is pushed to one side and released, it will oscillate back and forth. Under a heavier weight, it buckles to one side or the other.

This physical system may be crudely modeled by a dynamical system in the plane.

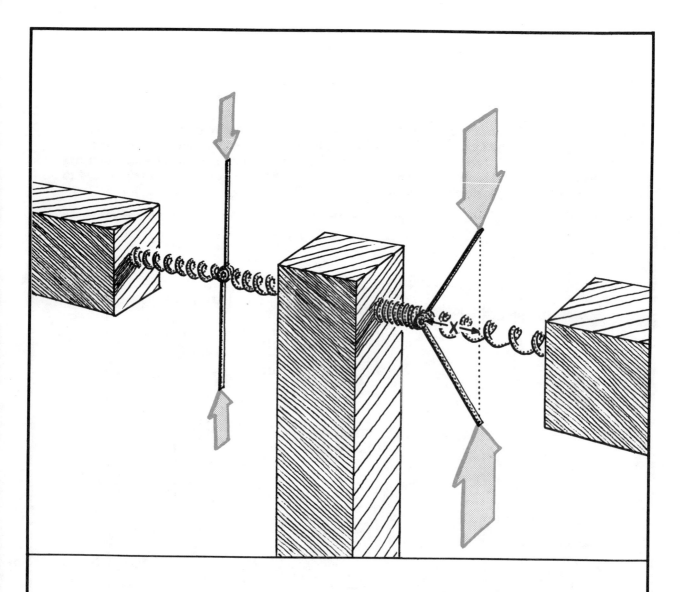

2.2.2. Stoker's model idealizes the elastic column as a hinged rod, restrained by two springs. When the compressive force is less than a critical value, the rods return to vertical alignment. Under a larger force, the hinge moves to one side or the other, compressing one of the springs and stretching the other. This corresponds to the buckling of the column.

The geometric model for the states of the hinged rod is the plane. The two observed parameters are the displacement of the hinge to the right of vertical, recorded on the horizontal axis: X, and the velocity of the hinge on the vertical axis: $V = X'$.

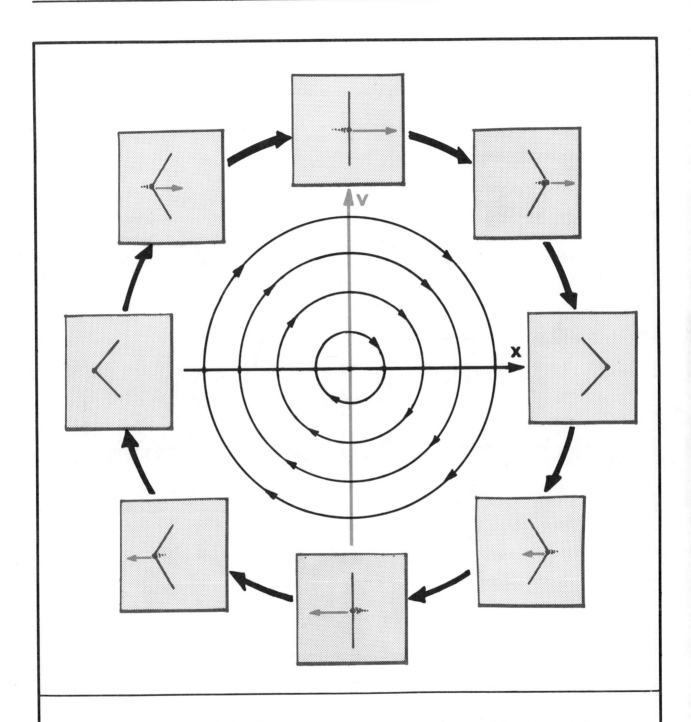

2.2.3. This tableau shows a succession of states, and their corresponding representations in the state space, as the hinge vibrates back and forth. Assume the weight is lighter than the critical value for buckling and there is no friction in the system. Then the phase portrait is a *center*, as in the preceding example. That is, it consists of concentric closed trajectories. These are periodic trajectories, representing oscillations. The breadth of a closed trajectory represents the *amplitude* of the oscillation.

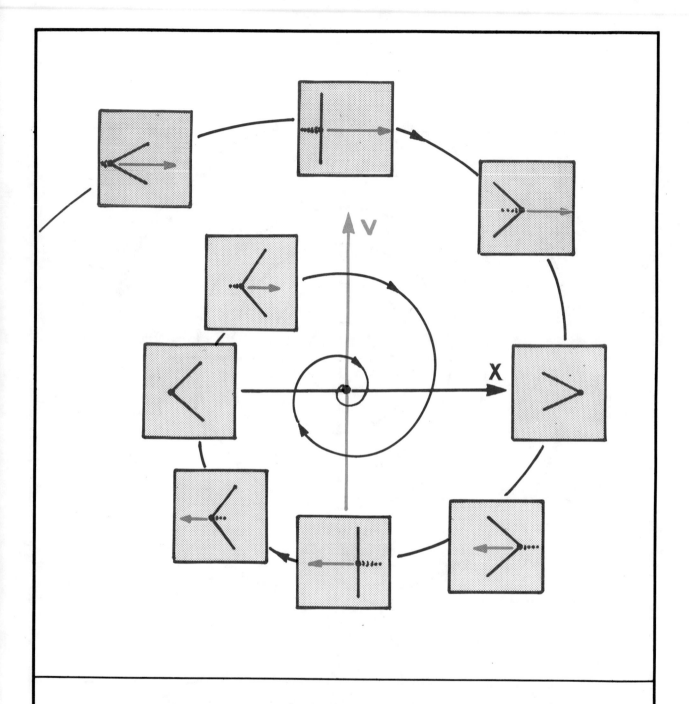

2.2.4.   Adding friction, the concentric loops are replaced by spiraling trajectories. As in the case of the pendulum with friction, they approach the origin, as the amplitude of oscillation decreases to zero. This is a typical phase portrait, with one basin, surrounding its unique *attractor* − in this case, a *focal point*.   Recall that a point attractor, or *rest point*, is a critical point (equilibrium point, limit point) which attracts all nearby initial states. Thus, all the state parameters describing these nearby initial states evolve asymptotically, as time increases, to constant values. Their omega-limit state is at rest, at the rest point.

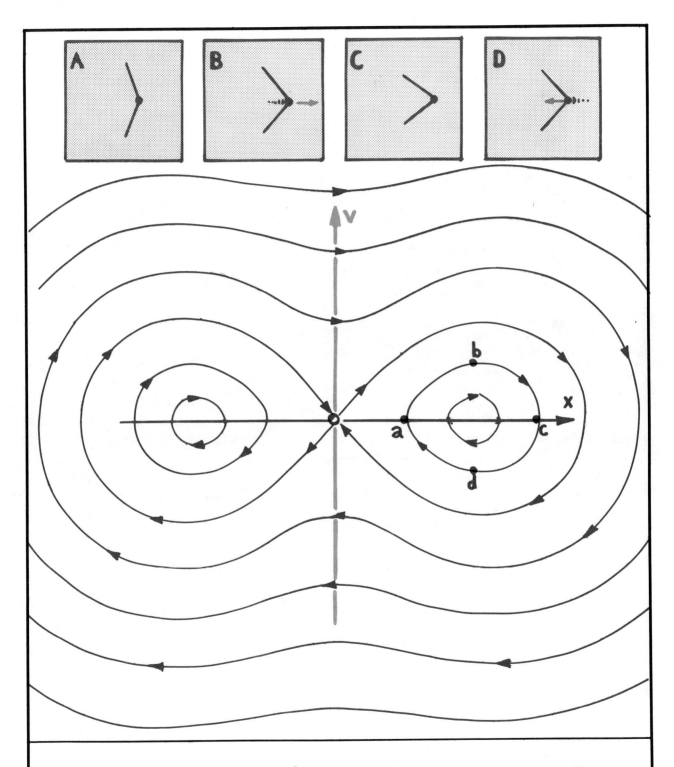

2.2.5. This is the phase portrait for the frictionless system, with the heavier weight. The sequence of states — A,B,C,D,A — describes an oscillation around an average displacement to the right. The *equilibrium point* at the origin is a *saddle.*

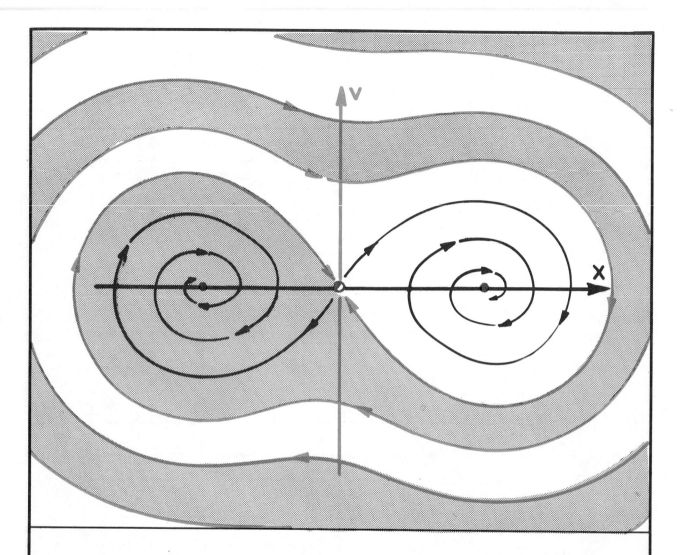

2.2.6. Adding friction, the concentric circles are again replaced by spirals. This is a typical portrait with two *basins*, a *focal point attractor* in each. Note that the insets of the saddle at the origin — shown here in green — do not belong to either basin. Thus, they comprise the *separatrix*, defining the boundaries of the two basins, one of which is shaded here. The relative area of a basin determines the *probability* of its attractor. That is, the chance of choosing an initial state which evolves to the attractor on the left is proportional to the area of the shaded basin. In this example, the two point attractors are equally probable.

This example is like the gradient systems described earlier, in that the limit sets are all points. Thanks to friction, closed trajectories are impossible. But unlike those of gradient systems, the trajectories approach the limit point in spirals, rather than radially. The point attractor is a focal point. The spirals correspond to oscillations of diminishing amplitude —*damped oscillations*. The point attractor which occurs in gradient systems is *radial*. That is, the approaching trajectories do not spiral. The radial type of point attractor is sometimes called a *star point*, or *node*.

2.3   PERCUSSION INSTRUMENTS

In *The Theory of Sound,* Lord Rayleigh studied separately the production, propagation, and reception of sound. His efforts to explain the production of sound by musical instruments became the theory of nonlinear oscillations[2]. From the point of view of dynamics, musical instruments may be divided in two classes:

—*percussion instruments,* such as drums, guitars, and pianos, to be modeled by damped oscillations (focal points, or point attractors of spiral type), and

—*sustained instruments,* such as bowed strings and winds, which are to be modeled by self-sustaining oscillations (periodic attractors).

This section describes the classical models for the percussion instruments. The sustained instruments will be treated in the next section.

2.3.1.   The percussion instruments all produce musical tones which decay (die out) in time. We hear the transient response of the system. The asymptotic limit of the audible transient is silence. Although in principle it takes a very long time for the note to die away, it actually becomes inaudible a short time after being struck.

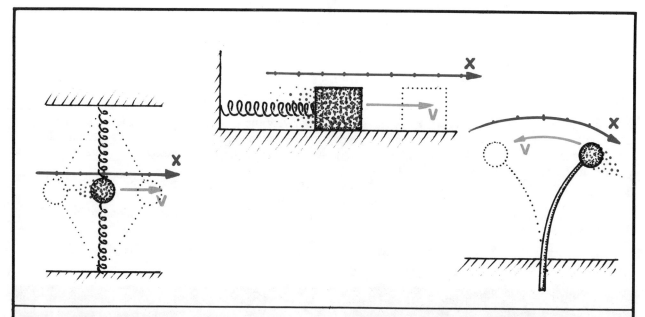

2.3.2. Simple mechanical models for the most sophisticated instruments look like an elementary physics lab. Different configurations of springs and weights behave, very approximately, like the instruments. As in the case of the buckling column, discussed in the preceding section, the resulting model is surprisingly useful.

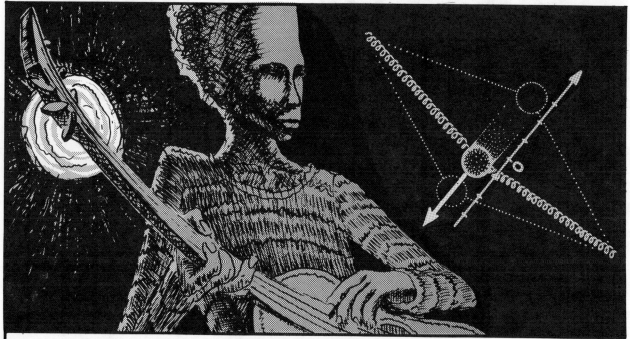

2.3.3. For example, the mechanical model for a plucked string is two linear springs of equal length, with a weight between them. The springs are stretched in-line, and the weight moves only along the line perpendicular to the springs.

2.3.4.   As in the two preceding applications, the geometric model for the
state space of the mechanical system is
the plane. The parameters are the displacement of the weight to the right of
equilibrium, and the velocity of its motion.   If there is no friction in the
mechanical model, the phase portrait
of its dynamical model (Newton's Law
of Motion) is a *center*. This is very
much like the frictionless pendulum,
described at the beginning of this
chapter.

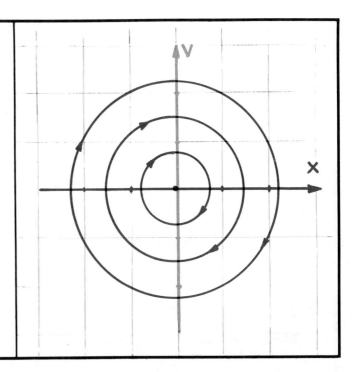

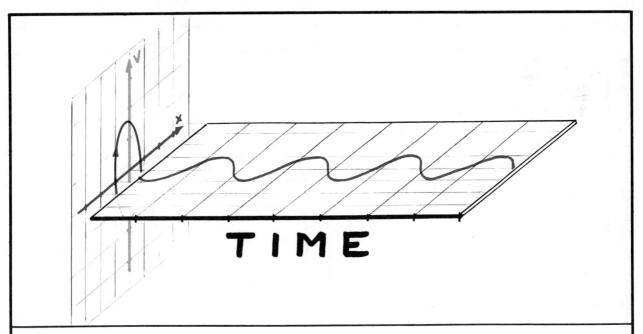

2.3.5.   Each trajectory of the center is closed. As described in Chapter 1, the time series
of a preferred parameter (for example, the displacement of the weight concentrated at the
center of the string) is a periodic function. The weight oscillates back and forth, periodically. In fact, under simplifying assumptions, this motion is sinusoidal. It corresponds to
a *pure tone*.

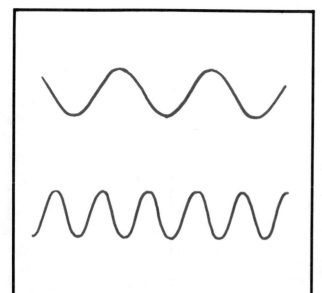

2.3.6. More oscillations per second correspond to higher frequencies, or tones.

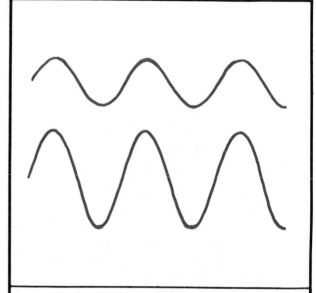

2.3.7. The vertical displacement of the time series, or *amplitude,* corresponds to the loudness of the note.

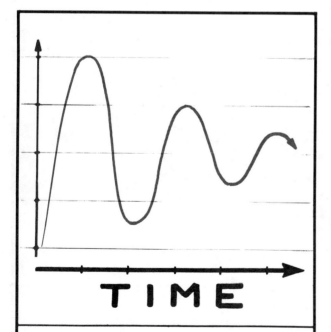

2.3.8. The time series of a plucked guitar or piano string is a function which decays, as shown here.

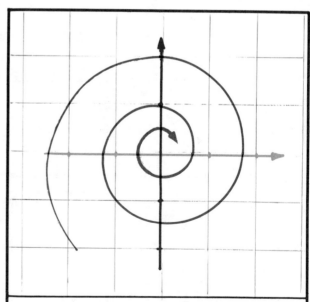

2.3.9. A trajectory in the state space of the mechanical model must spiral asymptotically toward the origin, in order to have the function on the left as its time series.

A model for decaying tones must include friction. To further simplify the discussion, we now replace the two springs (perpendicular to the motion of the weight) with a single spring (along the line of motion).

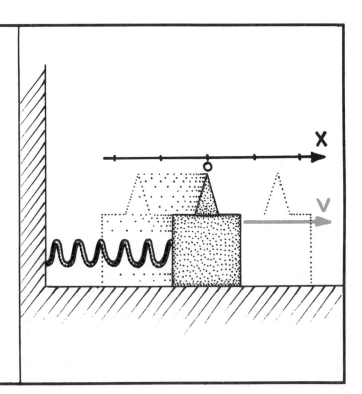

2.3.10. The dynamical model (Newton's Law) for this mechanical system is similar to that of the preceding system, with the two collinear springs. The difference is insignificant for small displacements.

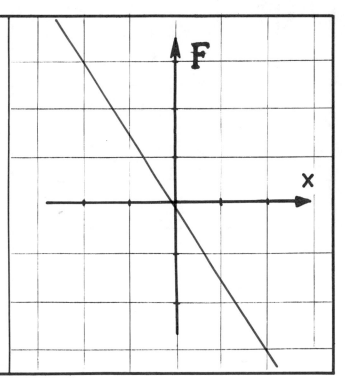

2.3.11. Next, in the single spring system, we assume the spring is linear. That is, we assume *Hooke's Law:* The force required to extend the spring a certain distance is a constant times that distance. Here is a graph of force versus extension, under this assumption.

The mechanical system described here, with a linear spring, is called the *harmonic oscillator*. Without going through the mathematical analysis of this system, which is classical, we simply present the results.

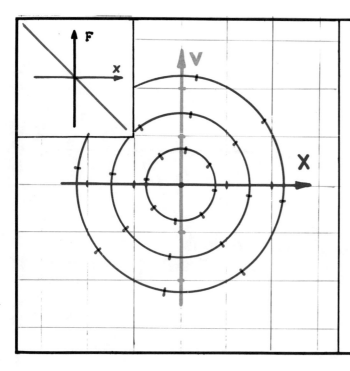

2.3.12. If after all there is no friction between the weight and the surface it moves upon, the phase portrait is a center. Further, all the concentric trajectories are periodic, with *the same period*. This is shown by the tick marks on the trajectories, in this illustration. Thus, no matter how hard you pluck the string, the note will have the same pitch. The spring characteristic (force versus extension) is shown in the inset.

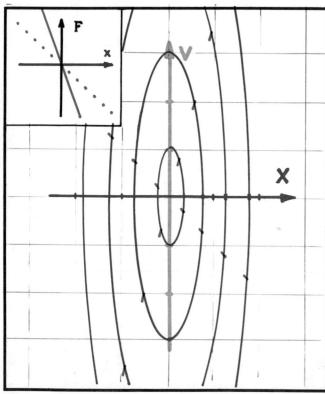

2.3.13. Tightening the guitar string corresponds to increasing the slope of the spring force versus extension graph. The steeper spring characteristic is again shown in the inset. The phase portrait is still a center, with concentric (more eccentric) elliptical trajectories. All these periodic trajectories still have the same period. But it is shorter than before. That is, the pitch (frequency) of the oscillation is greater.

The harmonic oscillator may be a poor model for a guitar string, for two reasons: (1) a guitar string is not linear, and (2) it is not frictionless either. Let's remove these objections one at a time.

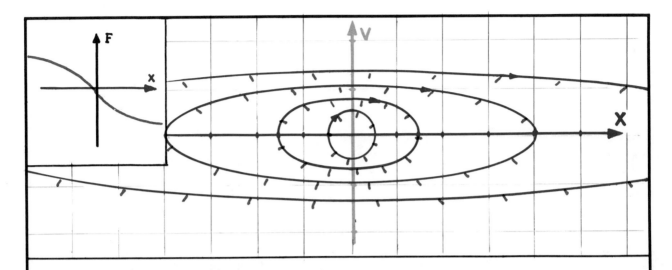

2.3.14. Suppose the spring is not linear. The graph of force versus extension will not be a line. But we may still suppose that its deviation from linearity is symmetric. The simplest such deviation from linearity is a *cubic* one. Two cases have been extensively studied. In one of these, called the *soft spring*, the force is less than linear, as shown in the inset. The phase portrait in this case is still a center. But in this case, the eccentricity (and thus also the frequency) of the periodic trajectories depends upon the amplitude. The larger trajectories have a lower frequency. Thus, the pitch of the plucked note will be lower for louder notes than for softer ones — such a string may make a poor guitar.

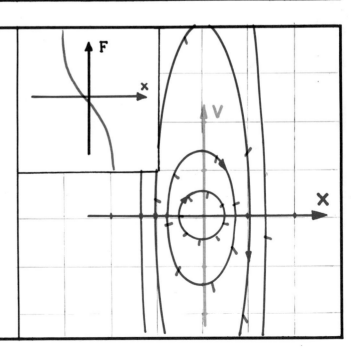

2.3.15. The other well-studied case of a nonlinear spring is the *hard spring*. Here, the spring force is a cubic function which is *more* than linear, as shown in the inset. The larger trajectories have a higher frequency in this case.

Now we deal with objection (2), by introducing friction between the weight and its supporting surface.

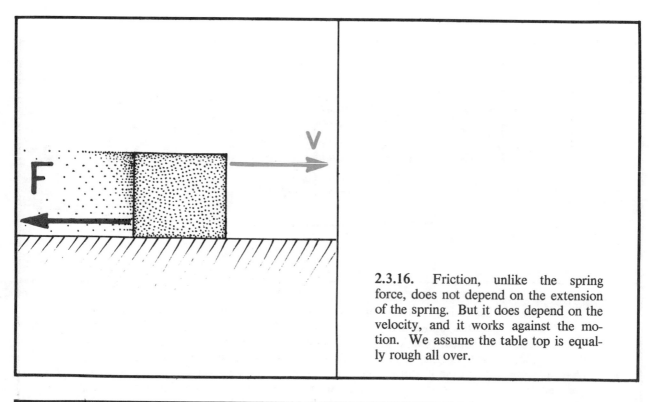

2.3.16. Friction, unlike the spring force, does not depend on the extension of the spring. But it does depend on the velocity, and it works against the motion. We assume the table top is equally rough all over.

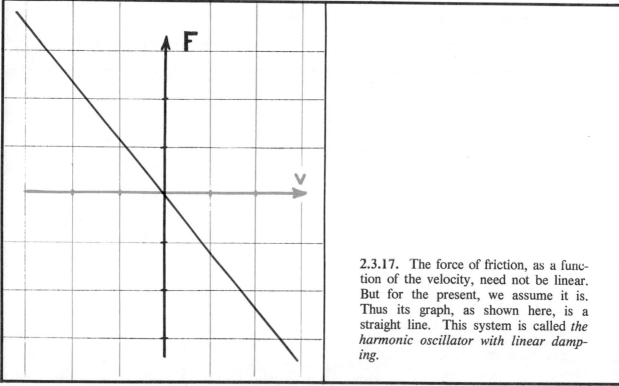

2.3.17. The force of friction, as a function of the velocity, need not be linear. But for the present, we assume it is. Thus its graph, as shown here, is a straight line. This system is called *the harmonic oscillator with linear damping.*

The mathematical analysis of the damped harmonic oscillator, likewise, is classical. Here again, we simply present the results of this analysis, as it was known to Newton. We return to the case of a linear spring, but add linear friction.

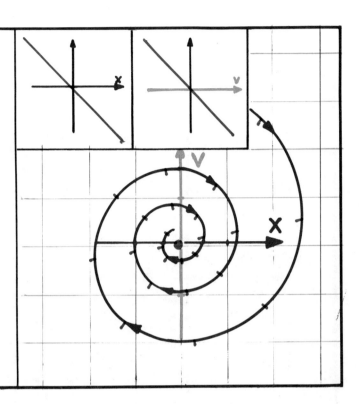

2.3.18. The phase portrait of the damped linear oscillator has a point attractor (focal point) at the origin. This is very like the damped pendulum. The linear spring and friction functions are shown in the insets. The damped oscillation has a constant period, (hence also, constant frequency or pitch) which is the same as the undamped system. The amplitude decays exponentially.

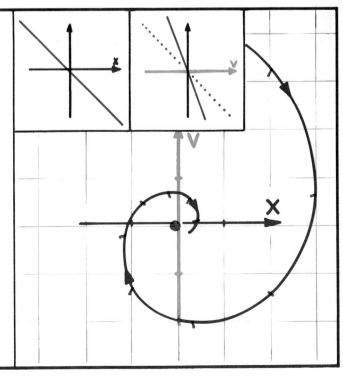

2.3.19. In the case of greater friction, the same spring will exhibit damped oscillation of the same period. But in this case, the amplitude decays faster. The spiraling trajectory approaches the focal point more quickly.

Now let's consider the spring model with both objections eliminated.

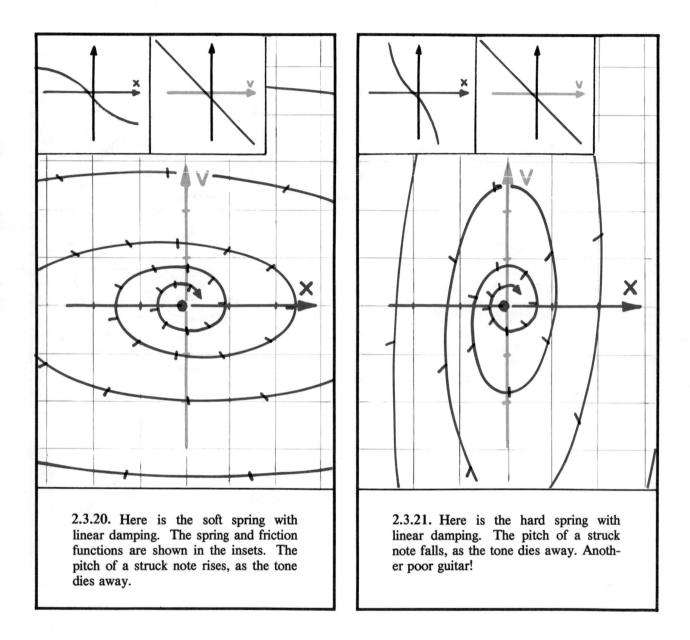

2.3.20. Here is the soft spring with linear damping. The spring and friction functions are shown in the insets. The pitch of a struck note rises, as the tone dies away.

2.3.21. Here is the hard spring with linear damping. The pitch of a struck note falls, as the tone dies away. Another poor guitar!

All the possible combinations and deviations from these simplifying assumptions have been explored, but by now you get the idea: A damped nonlinear oscillator is a reasonable model for a percussion instrument.

As a general rule, *attractors* model the observed states of the system.

Our first three applications violate this general rule. For in this case, the transient response models the tone heard, the attractor models the silence which follows. And in the preceding examples, the transient response modeled the damped oscillation of the pendulum or column, the attractor modeled the stillness which follows.

Finally, let's consider an extreme variation: what if the friction force were reversed?

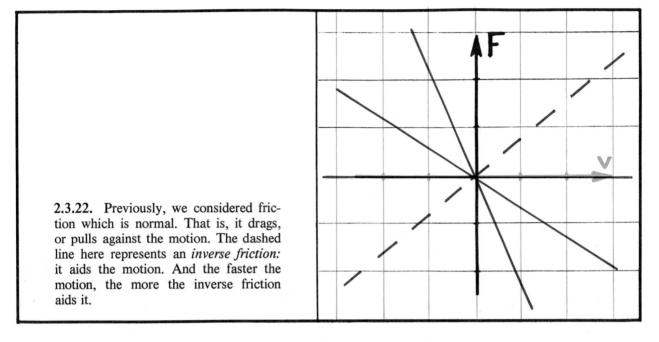

2.3.22. Previously, we considered friction which is normal. That is, it drags, or pulls against the motion. The dashed line here represents an *inverse friction:* it aids the motion. And the faster the motion, the more the inverse friction aids it.

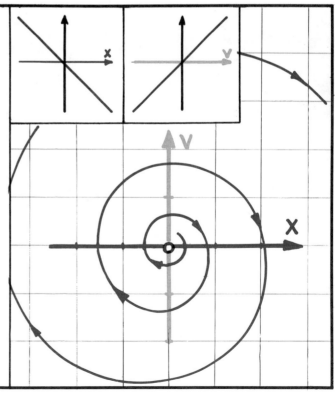

2.3.23. Here is the classical phase portrait for the harmonic oscillator with inverse friction. Again, the spring and friction graphs are shown in the insets. Here we have a *point repellor* at the origin. For any initial state, the alpha-limit set (asymptotic limit for the infinitely distant past) is the origin (no motion). The time series for any of these trajectories is an oscillation which grows (exponentially) with time. Presumably the spring breaks after a while.

In the next chapter, we will find this unusual model useful.

2.4 PREDATORS AND PREY

In this section, we illustrate an ecological application. This is a 1925 classic, due to Lotka and Volterra, the early pioneers of mathematical biology[3]. Consider a fictitious ecosystem, containing substantial populations of two species only— say big fish and small fry—along with a large supply of food for small fry . The choice of a state space for this application is easy.

The number of small fry and the number of big fish, respectively, are represented as coordinates in the plane.

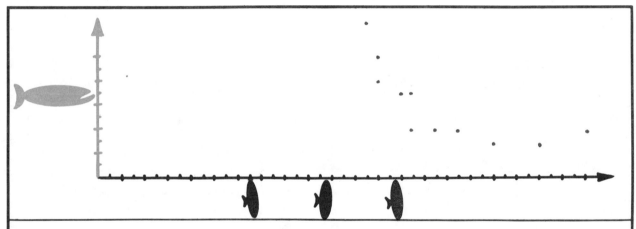

2.4.2. To apply the modeling concepts of dynamics, the dotted lines must be idealized into continuous curves by interpolation.

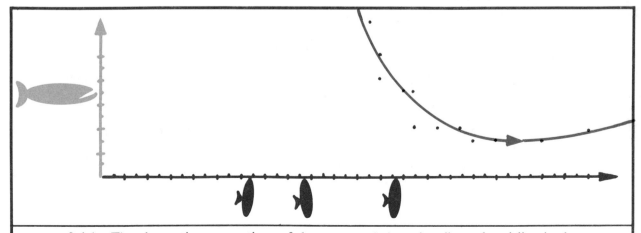

2.4.1. The observations, over time, of the two populations describe a dotted line in the plane. Births and deaths change the coordinates by integers, a few at a time.

The dynamical system for this model, the *Lotka-Volterra vectorfield,* **can be roughly described in four regions.**

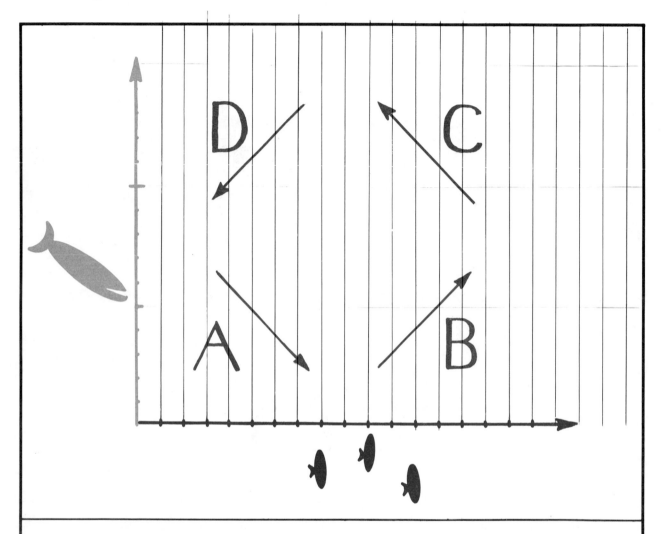

2.4.3. Region A. In this part of the state space, both populations are relatively low. When both populations are low, big fish decrease for lack of food (small fish) while small fry increase thanks to less predation. This is the habitual tendency for states in this region. The interpretation of this tendency as a bound velocity vector is shown here, in region A.

Region B. In this region, there are many small fry, but relatively few predators. But when there are many small fry and few big fish, both populations increase. This is interpreted by the direction of the vector shown in region B.

Region C. Here both populations are relatively large. The big fish are well fed and multiply, while the small fry population declines disastrously. This tendency is shown by the vector in region C.

Region D. In this part of the state space, there are few small fry but many big fish. Both populations decline. This tendency is shown by the vector in region D.

The Lotka-Volterra vectorfield is not just *some* vectorfield with these features, it is a *particular* one, which seemed the simplest choice at the time.

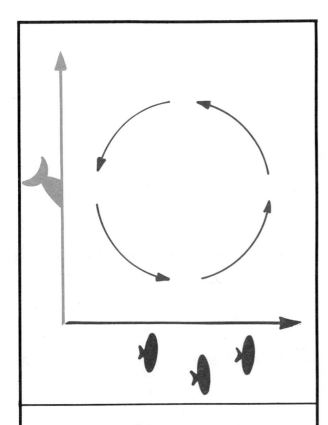

2.4.4. The phase portrait of this system can be visualized, in part, from these features: the flow tends to circulate counterclockwise. The ecologist would like to know what happens to the two populations *in the long run.*

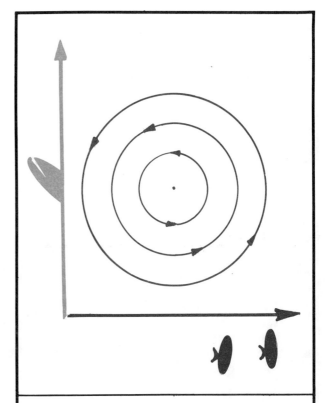

2.4.5. The answer, in this case, is a prediction of the mathematical model. The phase portrait is a *center:* a nest of closed trajectories, around a central equilibrium point[4].

Conclusion: every trajectory is periodic. Each initial population of big fish and small fry will recur periodically.

Now that the modeling process had been described, we may return to the question: why bother? This question has an exceptionally convincing answer, which accounts for the numerous examples of the process now proliferating in the literature of applied dynamics: *dynamical systems theory tells what to expect in the long run.* In this case, the two populations persistently oscillate. The same cycle of population numbers, for both species, will recur indefinitely, each time with the same elapsed time, or *period.* This is an example of a *prediction forever.* The periodicity of fish populations in the Adriatic actually inspired Volterra to make his model.

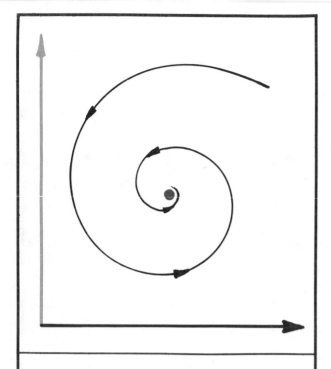

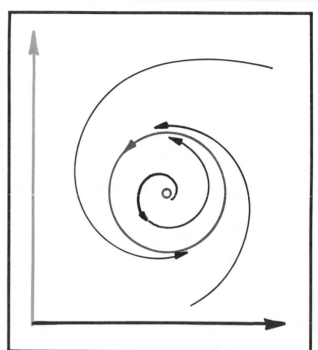

2.4.6. If some kind of ecological friction were added to the model, the center would become a focal point attractor. This would be a reasonable model for an ecological system in static equilibrium.

2.4.7. A more subtle modification of the model could result in a phase portrait like this, with only one periodic trajectory. This would be a more satisfying model for the observed periodicity of the fish in the Adriatic.

These improvements to the Volterra and Lotka model have been made recently[5]. But the type of phase portrait on the right, with the limit cycle, was well-known to Lord Rayleigh, as we shall see in the next Chapter.

3. VIBRATIONS: LIMIT CYCLES IN 2D FROM RAYLEIGH TO RASHEVSKY

The history of dynamics, from Pythagoras to the present, has been enlivened by music. Until around 1800, this history was dominated by the limit point concept. Then, Chladni's experiments with musical instruments attracted Napoleon's attention. And the limit cycle idea began to grow in the consciousness of the scientific community. This is an abstract analogue of the discovery of the wheel. In this chapter, we present the key dynamical steps of this bifurcation in the history of science.

3.1 WIND INSTRUMENTS

Following his analysis of the percussion instruments, Lord Rayleigh went further. In his attempt to explain all aspects of sound, he created a successful model for the sustained instruments. He managed to combine inverse friction to small motions with normal friction for large motions in a single dynamical system. The result is a simple example of *self-sustained oscillation*. The same model turned out, 45 years later, to be useful in the field of radio frequency electronics. This later application is described in the last chapter. Here, we resume our story in 1877.

3.1.1. What do these instruments have in common with a radio transmitter? As long as you don't run out of juice, they keep on playing.

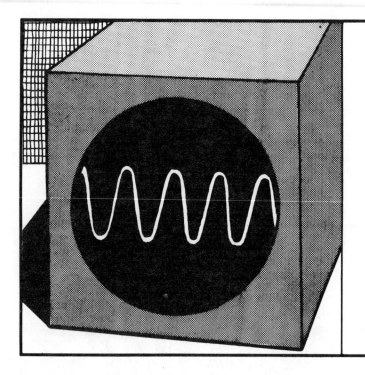

3.1.2. The sound of a sustained instrument, portrayed as a time series (amplitude of air motion versus time) by an oscilloscope for example, is a periodic function which does not decay in time. As long as the player puts energy into the instrument, the oscillation may be sustained at the same loudness (amplitude).

The dynamical model must have a closed trajectory, or periodic attractor, with this function as the time series for a preferred parameter. For the sake of definiteness, let's choose a clarinet reed as the target of the model.

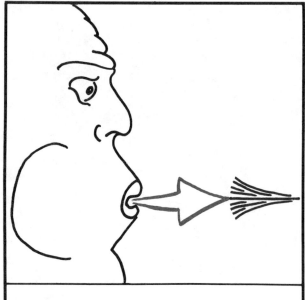

3.1.3. By blowing along the reed, the clarinetist adds energy to the system, sustaining the vibration.

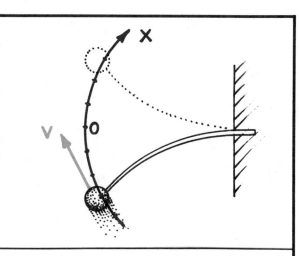

3.1.4. The 1870's style model for the reed is a flexible wand, with a concentrated small weight at the end. Somehow, we agree on a way to measure the amount the reed is bent, and its velocity (rate of change of this amount).

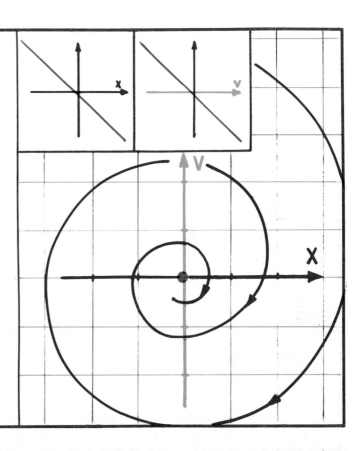

3.1.5. With no blowing, the wand is a type of pendulum. A reasonable dynamical model will look like this. A point attractor of spiral type is located at the origin. This is the spring model from the preceding section. The characteristic functions describing the damping and the spring are displayed in the insets.

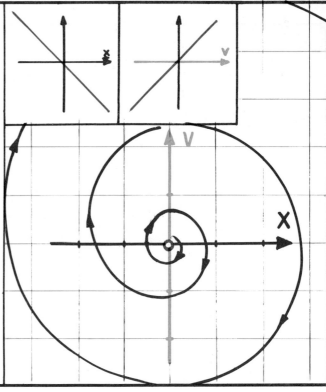

3.1.6. Rayleigh modifies the spring model to include the clarinetist by replacing the behavior near the origin by inverse friction. As described at the end of the preceding section, the origin becomes a point repellor. The behavior far from the origin is not changed.

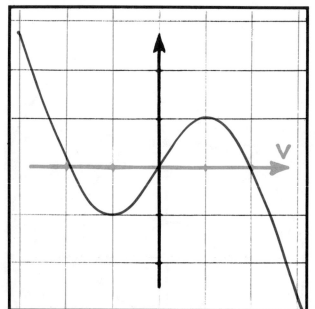

3.1.7. A simple way to mate normal friction for large motions with inverse friction for small motions is with this characteristic curve, a cubic (polynomial of degree three). This is the simplest curve with negative slope near the origin, and positive slope far away.

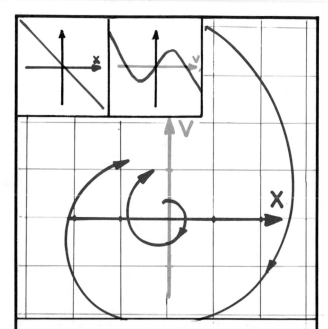

3.1.8. The resulting phase portrait has a point repellor at the origin. Yet far from the origin, all the trajectories are spiraling in.

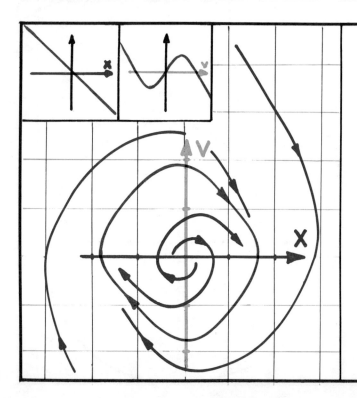

3.1.9. Between the distant spiraling in and the central spiraling out, a periodic trajectory (limit cycle) is trapped. Although this was obvious to Lord Rayleigh, mathematicians succeeded in proving it to their own satisfaction only about 50 years later.

This limit cycle is the dynamic model for the sustained oscillation of the blown clarinet reed. What is the relationship between the parameters in the model and the sound of the clarinet?

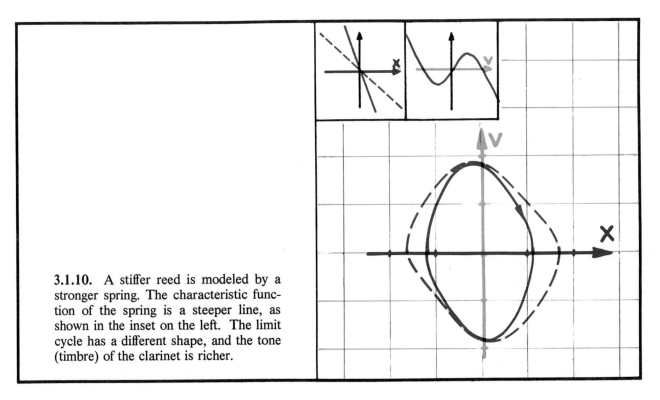

3.1.10. A stiffer reed is modeled by a stronger spring. The characteristic function of the spring is a steeper line, as shown in the inset on the left. The limit cycle has a different shape, and the tone (timbre) of the clarinet is richer.

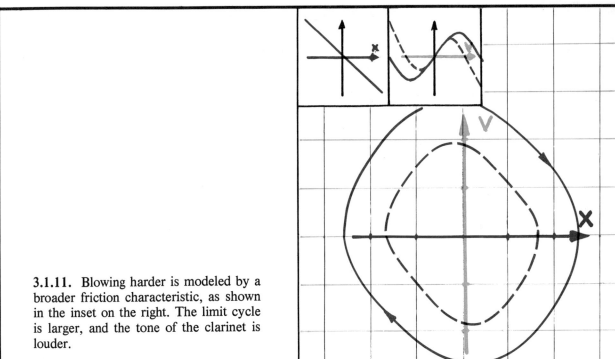

3.1.11. Blowing harder is modeled by a broader friction characteristic, as shown in the inset on the right. The limit cycle is larger, and the tone of the clarinet is louder.

This example of Lord Rayleigh's turned out to be the most important single item in the dynamics field for a century. To get more familiar with it, let's start again, this time, with a violin.

3.2   BOWED INSTRUMENTS

Many musical instruments produce sustained tones with the bowing mechanism. Besides the violin, cello, bass, and so on, Lord Rayleigh explicitly mentions the wine goblet, bowed by a finger on the rim. His great precursor, Chladni, applied his experimental bow to plates of glass, in hopes of creating new instruments.

To choose one example, let's consider the violin.

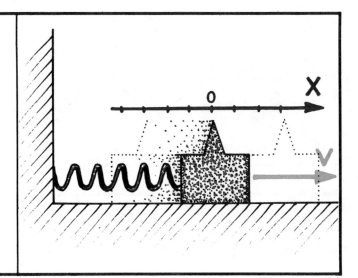

3.2.1. Ignoring the bow, the string of the violin is modeled by this 19th century gadget. This is identical to the guitar string model of the preceding section. Again, we use x to denote values of the displacement of the spring (that is, the string at the point of bowing) and v for the velocity, that is, the rate of change of the displacement.

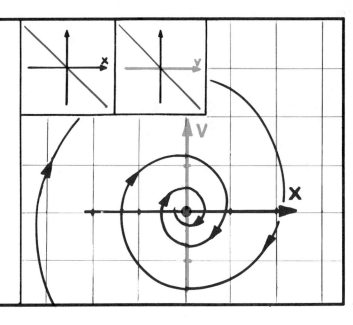

3.2.2. The phase portrait has, again, a *focal point attractor* at the origin. The *frequency* (rate of spiraling) and the rate of decay are determined by the characteristic functions of the spring (left inset) and the friction (right inset). Both of these are aspects of the violin string itself, and of its tightness.

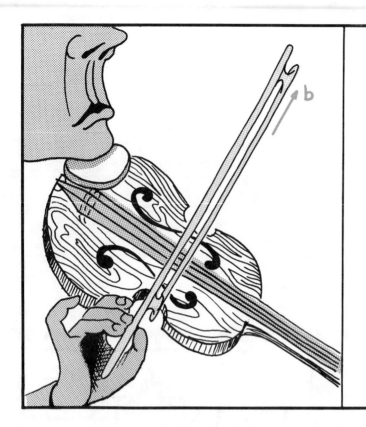

3.2.3. The violinist sustains the vibration by putting energy into the string with the bow. The friction of the bow on the string depends on the rate of bowing. We introduce a new symbol, *b,* to denote the rate of drawing the bow across the string.

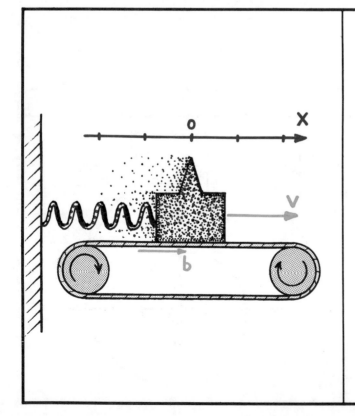

3.2.4. The spring model may be simply modified to include the action of the bow. Replace the tabletop on which the spring slides by a conveyor belt. This represents the bow. The weight, as before, represents the violin string.

The damping depends on the string only. It is not changed by the bow. The violinist uses rosin on the bow, making it sticky. This changes the shape of the curve describing the friction as a function of the velocity. To get the idea, we consider a weight moving on a sticky tabletop.

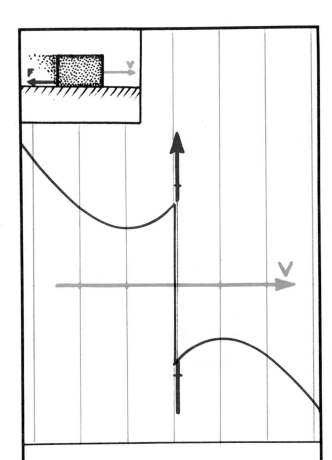

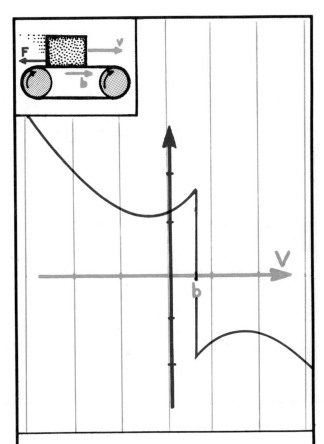

3.2.5. Further, suppose the tabletop is very sticky with rosin, and we just begin to push the weight lightly to the right. The speed is zero. But the force of friction is building up as we push. Suddenly it slides, as the force reaches a critical value. The same thing happens if we pull instead of push. This experimental situation is represented in the mechanical model by the shape of the characteristic function of friction shown here.

3.2.6. In the preceding discussion, we assumed the weight had hardly begun to move. Now, lets imagine the same experience — sticking, pulling up to critical force, then slipping — while the tabletop is moving relative to the weight, with speed, b. This situation is modeled by this friction function, obtained from the preceding example by sliding the graph to the right. You see, the tabletop is moving at speed, b. So when the weight sticks to the tabletop, it must be moving at the same speed, $v=b$. Thus, the vertical segment of the graph must be located at the common velocity.

Identifying the tabletop with the bow, and the spring-weight system with the violin string, we now have a dynamical model for the bowed string. The dynamical model corresponds to a 19th century mechanical model: a weight, fixed to a linear damped spring, oscillating on a conveyor belt. This dynamical model is the same as that for the clarinet reed, except that the smooth (cubic) friction function for the blown reed is replaced by this one with a glitch. The glitch is located at the speed of the bow. We have only drawn the function for one bowing speed. This happens to be positive. That is, the violinist is pushing toward positive deflection. We do not have, in this case, a precise function in mind. We just assume that it is shaped something like this.

Assuming the friction function characterizing the bowed violin string looks something like this, what can we deduce about the phase portrait of the dynamical system? (If you just want a quick answer to this question, you could skip to the end of the section.)

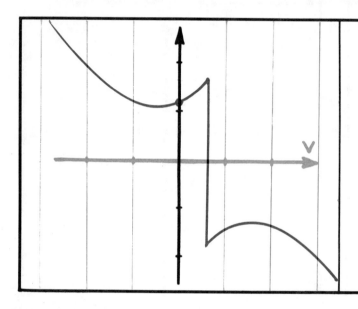

3.2.7.   When the motion pauses, the velocity is zero. The force of friction, read from the graph here, is then some number, $F(0)$. This friction is independent of the displacement of the spring. It depends only on the velocity.

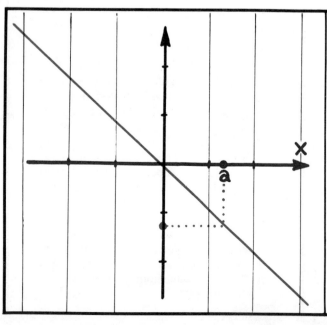

3.2.8.   Meanwhile, the force of the spring depends linearly on the deflection. It is independent of the velocity. For some certain deflection, $x=a$, the spring force will be $-F(0)$.

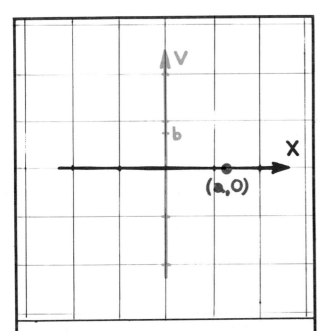

3.2.9. At this special deflection, and at zero velocity, the friction force exactly balances the spring force. The phase portrait has a critical point at *(a,0)*.

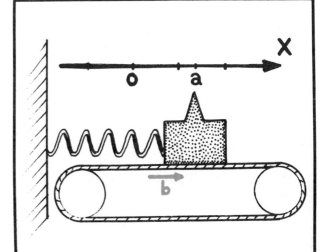

3.2.10. Here is the mechanical model at the critical point. The bow is moving to the right at its fixed speed, *b*. The weight has paused *(v=0)* at the critical displacement *(x=a)* where the friction force is balanced by the spring force.

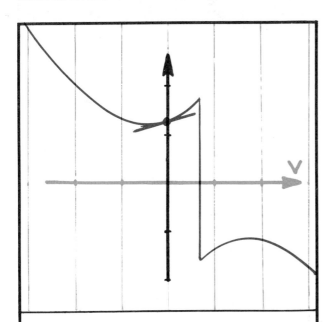

3.2.11. At this critical point, the inclination of the friction function is positive. This is the situation called *inverse friction* at the end of Section 2.3.

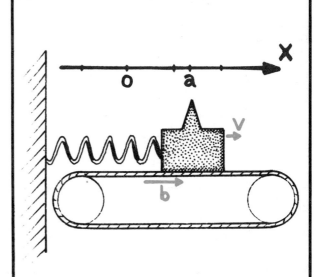

3.2.12. This means that a small motion of the weight to either side will create a runaway oscillation, as explained at the end of Section 2.3.

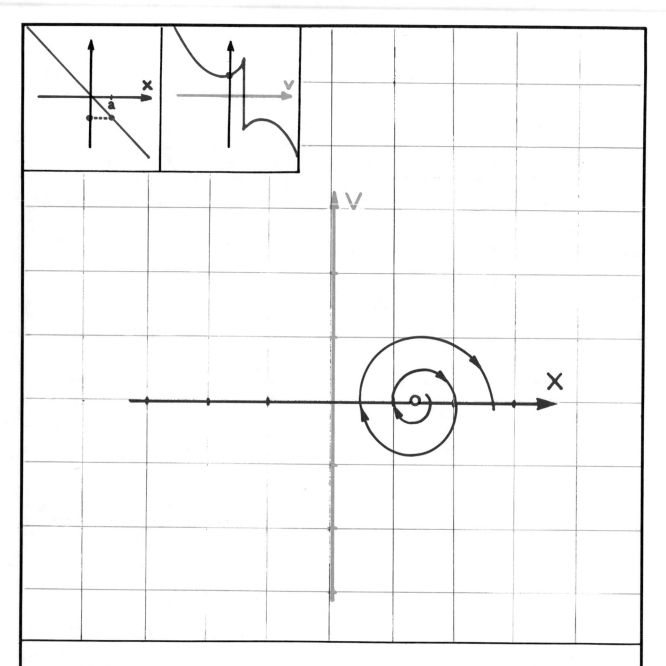

3.2.13. So far, we have figured out that there is a critical point in the phase portrait of our dynamical model for the bowed violin string, and that it is a *focal point repellor*.

As the small motions about the critical point grow, and the large motions decay, it seems plausible that there is a limit cycle in the phase portrait for the bowed violin string. This would be just like the model for the blown clarinet reed. Unfortunately, this cannot be proven without assuming more about the shape of the friction function.

Assuming that there is a limit cycle in the phase portrait, what does this mean the violin string is doing when it is bowed? It means the endless cycle of states shown here: 1,2,3,4,1,2,3,4, and so on. *Note:* The spring force graphs are drawn upside down here for easier comparison with the adjacent friction force graphs in the inserts.

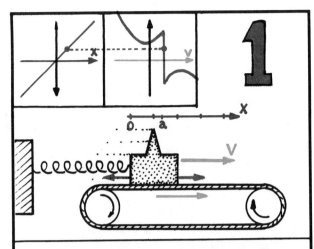

3.2.14. 1. STUCK. The friction and spring forces are balanced, and the weight is stuck to the belt. The weight is a little to the right of zero (equilibrium of the spring) but not as far as the critical point.

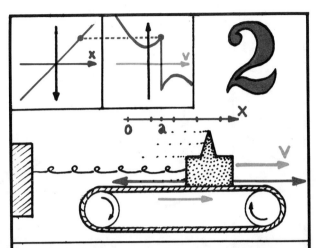

3.2.15. 2. BEGINNING TO SLIP. The friction and spring forces are balanced, but larger, as the displacement increases. At the critical force for friction, slipping begins.

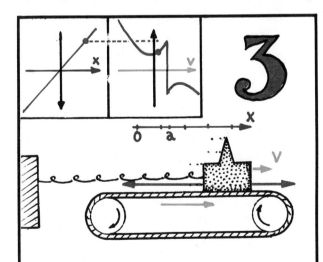

3.2.16 3. SLIPPING. When the velocity begins decreasing, while x is still increasing, then the sudden drop in the friction yields the tug of war to the spring. The acceleration is negative. Rightward motion slows to a halt, the weight begins to move back to the left.

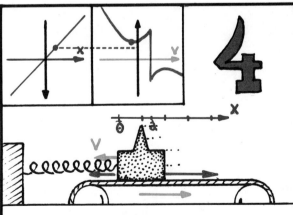

3.2.17. 4. GRABBING. When the leftward motion has decreased the spring force to a value smaller than the slipping friction, the tug of the belt wins once more. Motion to the left slows, the weight turns and begins once again to move to the right. When the velocity reaches the critical value (the red dot on the glitch) in the friction function, slippage is going to happen again. (return to 1).

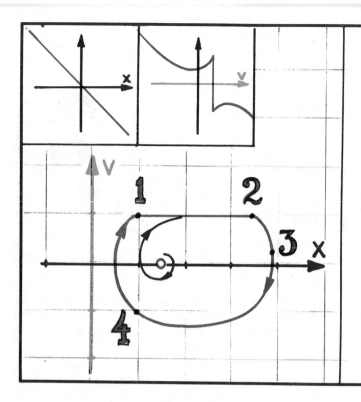

3.2.18. Here are the four stages in the cycle, located in the phase portrait. The rest of the limit cycle has been interpolated. The flat part at the top corresponds to the stuck phase. The repelling critical point is inside the cycle. Initial states near this repellor will spiral outward, clockwise, approaching the self-sustaining oscillation.

What happens outside the cycle? Let's make a change of scale, so the cycle is only about one tenth its former size.

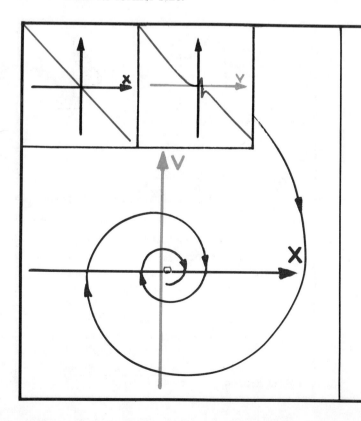

3.2.19. Now the inverse friction region is smaller than a grape seed, and is almost at the origin. The friction function looks essentially linear, and normally dissipative. The phase portrait has, roughly, an attractive point near the origin. Actually, it is not a point. It is an attractive region about the size of a grape seed or so, containing a limit cycle. The initial states in this picture, corresponding to very large scale motions of the weight on the conveyor belt, will decay to the vicinity of the limit cycle (the seed). So essentially, the closed trajectory is attractive (a limit cycle), as we have assumed all along.

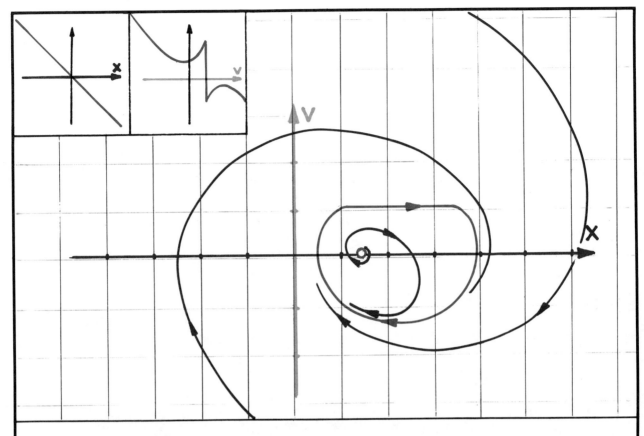

3.2.20. Returning to normal scale, here is the complete phase portrait for the dynamic model for the bowed violin string.

All this was child's play for Lord Rayleigh. He managed to make mechanical models very simply, learn from them, translate back and forth to the symbolic expressions of differential equations, and relate them equally well to electrical models. We will refer to the dynamical system used by him as a model for self-sustained oscillations —equally for wind or bowed instruments, or the electrical oscillator of Helmholtz —as *Rayleigh's system*. It comes up again in the next section, as *Van der Pol's model for electronic oscillations*, and again in Chapter 5.

3.3. RADIO TRANSMITTERS

Rayleigh had already observed that his model for self-sustained mechanical oscillations applied equally well to an electrical oscillator suggested a few years earlier by Helmholtz. Among Rayleigh's followers the early experimentalists, Duffing and Van der Pol were particularly influential. Duffing was especially interested in mechanical vibrations, while Van der Pol worked with the first electronic oscillators based on vacuum tubes. In the next two chapters, we will describe the main results of these two experimentalists. At this point, the work of Van der Pol provides a second example of the representation of an oscillating physical system by a dynamical model with a periodic attractor.

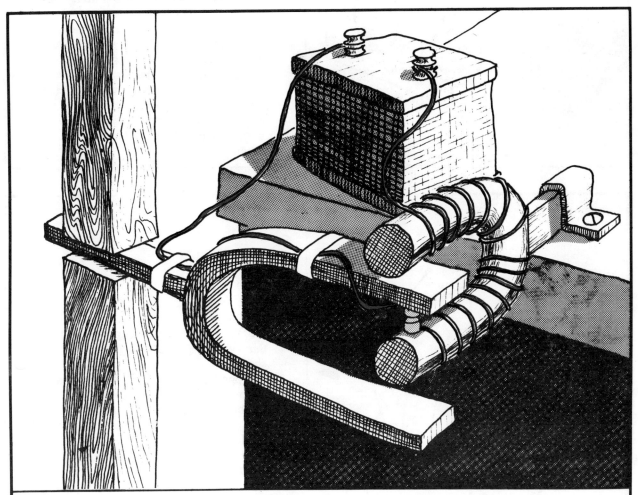

3.3.1. Here is the scheme of Helmholtz' electrical vibrator, the *tuning fork interrupter*. This device lives on, even today, as an alarm-bell ringer, or door-bell buzzer.

The invention of the triode vacuum tube made possible the realization of Helmholtz' scheme at very high frequencies, and so radio transmission was born. But to van der Pol, this device became an extremely manageable laboratory instrument for experimental dynamics.

3.3.2. The physical system consists of the original radio transmitter. The chassis contains power supplies, a triode vacuum tube, a *tank circuit* consisting of an inductive coil and a variable capacitor in parallel, load resistors, and a feedback coil from the plate tank circuit to the grid of the tube, to induce oscillations. The two dials on the front of the chassis monitor the radio frequency current and voltage at the plate of the tube.

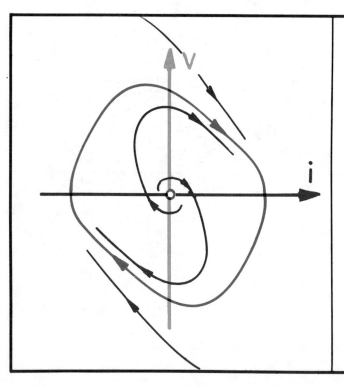

3.3.3. The observed parameters of this system are the voltage and current, shown by the panel meters. Thus, the appropriate state space is the plane. Here is the phase portrait of the vectorfield deduced by van der Pol for this system. This is based upon electronic circuit theory, now standard. Its chief features are a repelling equilibrium point at the origin, and a periodic attractor around the origin. The mathematical proof of these facts is arduous, compared to the ease of their discovery by experiments.

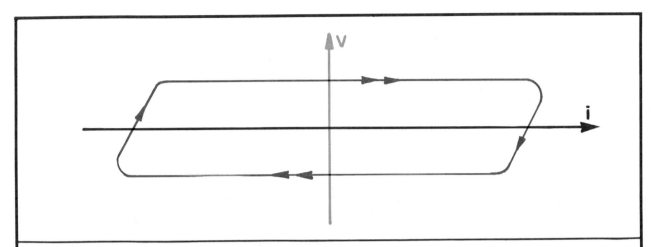

3.3.4. A simple modification of the preceding dynamical model for the triode oscillator yields this portrait, called a *relaxation oscillator*. The speed along the periodic attractor in this case is relatively slow on the near-vertical segments, and fast on the longer horizontal segments. Thus, the equilibrium oscillation lingers long at the minimum voltage (horizontal axis), snaps over to the maximum voltage, lingers there, then snaps back. And van der Pol himself proposed this as a model for the heart beat.

3.4   BIOLOGICAL MORPHOGENESIS

Many novel and exciting applications of dynamics to topics in biology and social theory were envisioned by Rashevsky. The best known of these, a model for biological morphogenesis, was rediscovered by Turing, and later studied by others. In this section, we will illustrate our interpretation of Rashevsky's model in the context of *phyllotaxis,* the morphogenesis of plant growth[3].

The empirical domain for this application is an idealized vine.

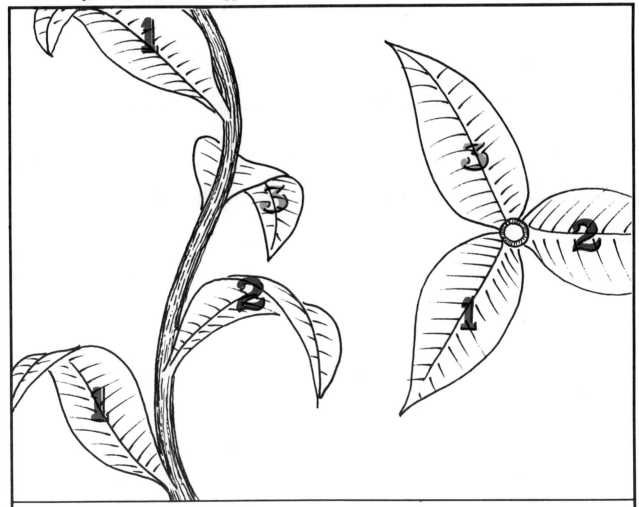

3.4.1.   Have you noticed how some plant stalks sprout branchlets symmetrically? This vine, for example, sprouts one branchlet at a time. The direction of these branchlets rotate around the stalk with trihedral symmetry.

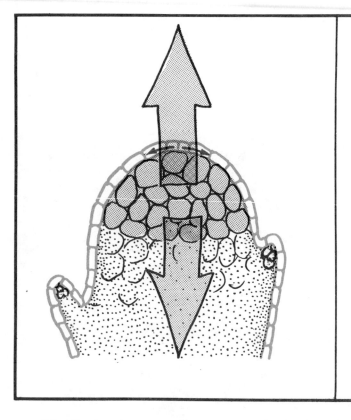

3.4.2. At the tip of the growing stalk is a growth bud. Beneath the epidermis is the *apical meristem,* a mass of undifferentiated, totipotent, cells. Trailing in their wake, as they move upward on the growing stalk, are various derivative, cytologically differentiated, cells. Among these are the leaf bud cells, the branch cells, and so on. The question of morphogenesis, also called *phyllotaxis* in this context, is the formation of the *pattern* of these differentiated cells, and thus, of the leaf buds and branchlets.

The **Rashevsky model** for morphogenesis is based on a ring of growth cells, around the circumference of the stalk, near the growth bud at the top. Let's build up his model for this ring of cells, one cell at a time.

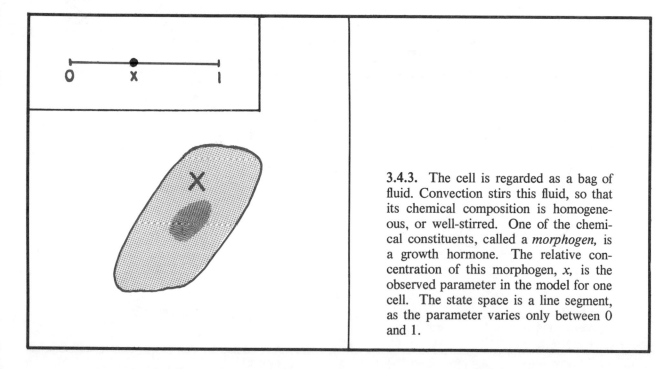

3.4.3. The cell is regarded as a bag of fluid. Convection stirs this fluid, so that its chemical composition is homogeneous, or well-stirred. One of the chemical constituents, called a *morphogen,* is a growth hormone. The relative concentration of this morphogen, $x,$ is the observed parameter in the model for one cell. The state space is a line segment, as the parameter varies only between 0 and 1.

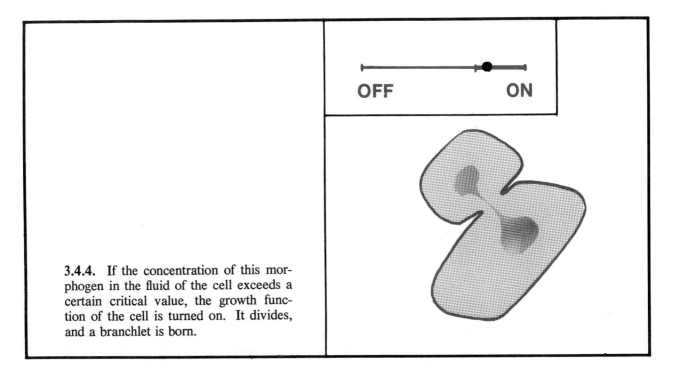

3.4.4. If the concentration of this morphogen in the fluid of the cell exceeds a certain critical value, the growth function of the cell is turned on. It divides, and a branchlet is born.

Next step: two cells, with one morphogen, in an *open system*. This means that the morphogen can come and go between the two-celled system and its environment.

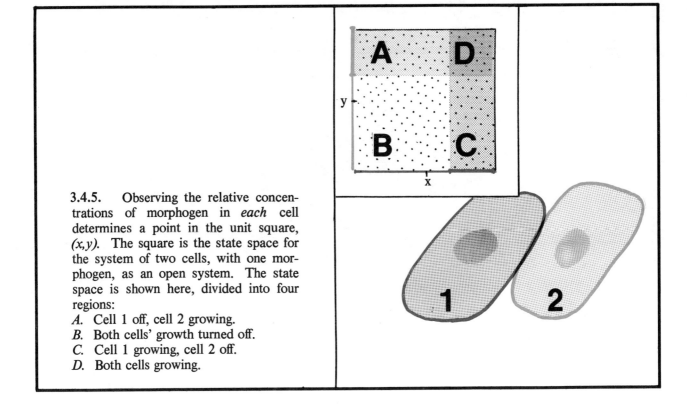

3.4.5. Observing the relative concentrations of morphogen in *each* cell determines a point in the unit square, *(x,y)*. The square is the state space for the system of two cells, with one morphogen, as an open system. The state space is shown here, divided into four regions:

A. Cell 1 off, cell 2 growing.
B. Both cells' growth turned off.
C. Cell 1 growing, cell 2 off.
D. Both cells growing.

Now let's close the system. In the closed system of two cells, the morphogen cannot enter or leave. The total amount is constant.

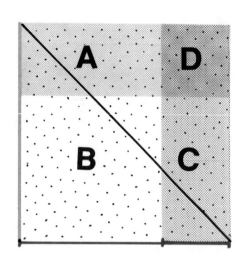

3.4.6. The state space for the closed system is a *subspace* of the square. Only the points of the square on the black line segment satisfy the constraint of the closed system: the sum of the concentrations is constant, or $x+y=1$.

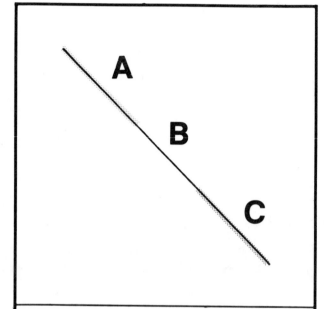

3.4.7. Extracting this line segment, we have the state space for the closed system. Notice that the line meets the zones *A, B,* and *C* of the square.

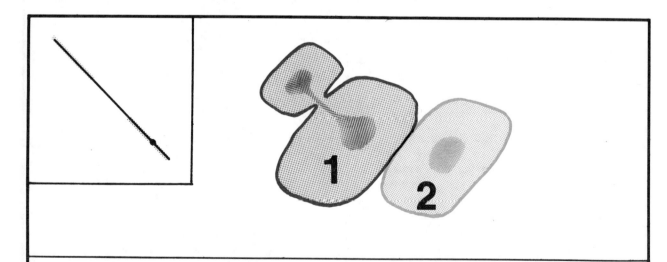

3.4.8. If the state of the closed system of two cells is in the line segment *A*, cell 1 is off, and cell 2 is growing. In segment *B*, neither cell is growing, and in segment *C*, cell 1 is growing, while cell 2 is off.

Final step: three cells, one morphogen, closed system. We think of them as a *ring* of cells.

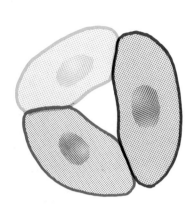

3.4.9. Here is the ring of three cells, with a uniform concentration of the morphogen in each. The point *(x,y,z)* in the unit cube of three-space represents a state of the system.

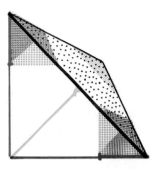

3.4.10. Plotting the state in three space, we find the assumption of a closed system, *x+y+z=1,* constrains the state to lie on this triangle, called the *unit simplex* of three space. This planar, equilateral triangle is the state space for the closed system of three cells, with one morphogen.

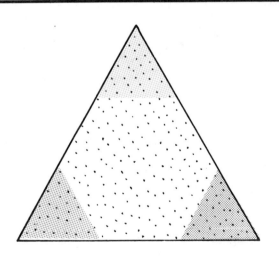

3.4.11. The three small triangles at the corner points of the state space correspond to a distribution of morphogen in which the amount in one of the cells exceeds the critical value for growth.

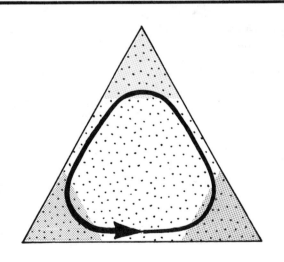

3.4.12. Suppose now that a dynamical system has been added to the model, and that it has a periodic attractor like this. Then periodically, one after another of the three cells is turned on, then off.

We may now connect with the bean stalk, by imagining a *stack* of rings of cells, as a simplified model for the *stalk*.

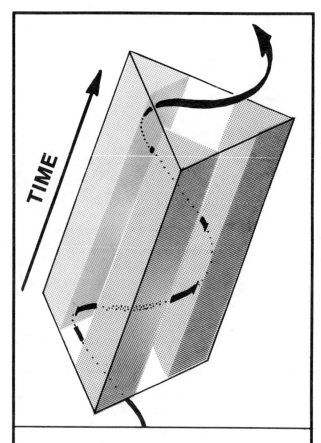

3.4.13. Here is the stack of rings of cells, each ring represented as an identical copy of the triangular model. Growth of the stalk upwards in time is represented by associating time with the upward direction. The periodic attractor of the preceding illustration thus gives rise to a *periodic time series* spiraling upwards in time.

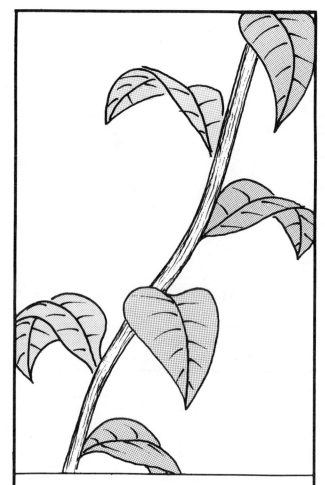

3.4.14. Successively, the turned on cell proceeds around the stalk, initiating branchlets. Many improvements to this model come immediately to mind, and no doubt they had occurred to Rashevsky already, in 1940.

The further biological and social applications of this scheme for morphogenesis await us, in the near future.

4. FORCED VIBRATIONS: Limit cycles in 3D from Rayleigh to Duffing

Lord Rayleigh's study of musical instruments provided the early examples of limit points and limit cycles in the plane, discussed in the two preceding chapters. He went on to study *forced oscillations,* with musical applications in mind. Besides these applications, involving tuning forks and the determination of pitch, he envisioned further applications to tides and electrical motors. This led him into experimental work, progressive abstraction into theory, and the foundation of a new branch of dynamics: *forced vibration.*

Forced vibration is one of the most significant topics in dynamics, and its potential applications are manifold. We distinguish two separate cases:

(1) A system which tends to rest is subject to a periodic force. Classical example: effect of mechanical vibration on a pendulum. Biological example: effect of the seasons on big fish and small fry.

(2) A system which tends to self-sustained oscillation is subject to a periodic force. The preceding section on biological morphogenesis, for example, suggests the question: what happens if a biological oscillator is influenced by an external periodic force, such as sunlight?

In the next chapter, we will describe the results obtained for case (2) by Rayleigh and van der Pol. In this chapter, we describe Rayleigh's work on case (1) with the double pendulum, and the related results obtained later by Duffing. We begin by constructing a three-dimensional model for the states of the forced system, the *ring model.*

4.1   THE RING MODEL FOR FORCED SPRINGS

To the early experimentalists, a self-sustained oscillator was hard to arrange, so they approximated one with a very large pendulum. The decay in the amplitude of its swing would be insignificant in a short experiment. As the source of the periodic force applied to the *driven system,* a smaller pendulum, it would be relatively unmoved by the motion of the driven system. Of course, we would properly consider this a *coupled system* of two swinging pendula. It only *approximates a forced vibration,* which is an unreal idealization. This was well understood by Rayleigh, who wrote[1]:

> As has already been stated, the distinction of forced and free vibrations is important; but it may be remarked that most of the forced vibrations which we shall have to consider as affecting a system, take their origin ultimately in the motion of a second system, which influences the first, and is influenced by it. A vibration may thus have to be reckoned as forced in its relation to a system whose limits are fixed arbitrarily, even when that system has a share in determining the period of the force which acts upon it. On a wider view of the matter embracing both the systems, the vibration in question will be recognized as free.

Our goal in this section is to turn this intuition into a geometric model in three dimensions. This is the state space for the coupled system. In it, a free vibration of the coupled system (equivalent to a forced oscillation of the driven system) is represented as a *periodic attractor.*

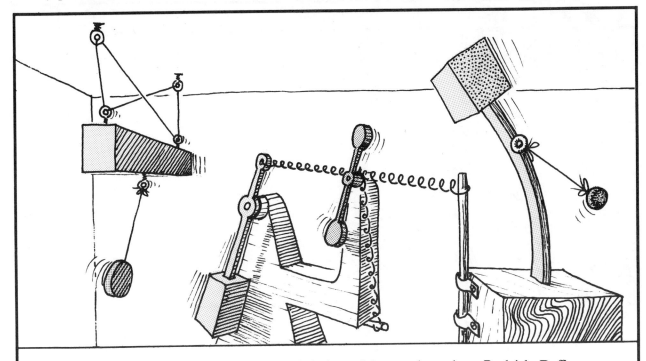

4.1.1. Here are the actual experimental devices of three early workers: Rayleigh, Duffing, and Ludeke.

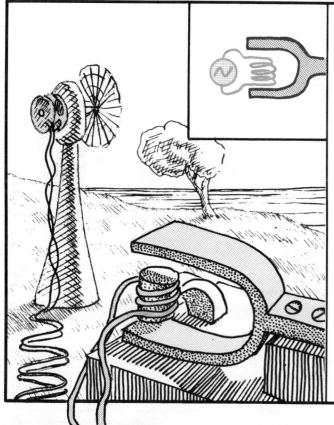

4.1.2. Helmholtz and Rayleigh also analyzed mechanical systems subject to periodic electrical forces. These systems more closely approximate the ideal forced vibration, in that the forcing system (alternating current generator) is relatively indifferent to the motion of the driven system (tuning fork).

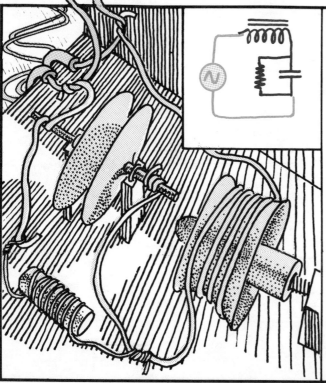

4.1.3. The early experimentalists also analyzed electrical systems subject to periodic electrical forces. For example, a parallel plate capacitor and an inductive coil in series was regarded as an electrical analogue of the tuning fork, in the driven system.

We now describe **Duffing's results, in the original mechanical context, the double pendulum. First, the driven system.**

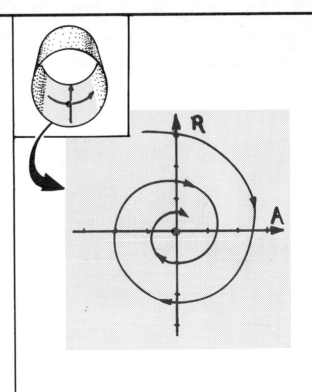

4.1.4. This pendulum will be the driven system in the mechanical apparatus for studying forced vibration in case (1). One observed parameter, *A,* is the angle of deflection of the bob from vertical. We also observe another parameter, the rate of change of the angle, *R.*

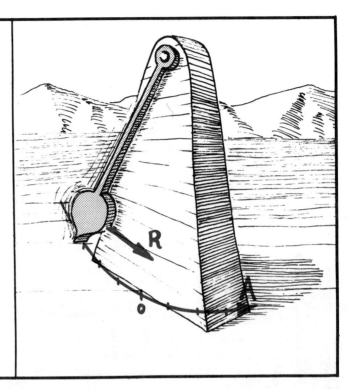

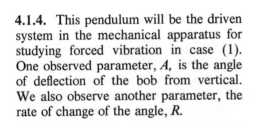

4.1.5. Recall that the state space for this device is a cylinder, as described in Section 2.1. But we may cut the cylinder and unroll it into a plane. This will be useful here, as we will consider only small motions of the bob, represented by points near the origin, *(A,R)=(0,0),* of the state space. Recall also that the phase portrait for this system has a focal point attractor at the origin.

And now, the driving system, with 20th century additions....

4.1.6. This turntable motor has a sophisticated governor, which will work hard to maintain a constant frequency of rotation. To its turntable is connected a push-rod and lever. The upper (pointed) end of this lever will eventually be the oscillating point of support for the driven pendulum.

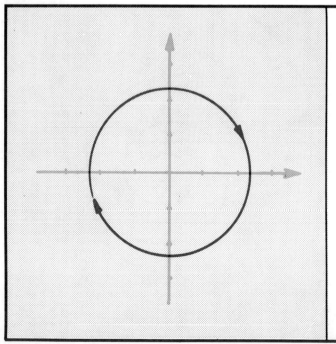

4.1.7. This motor is a replacement for the giant pendulum in Rayleigh's original scheme. It has a self-sustained oscillation. Like the clarinet reed and violin string of the preceding chapter, its dynamical model has an attractive limit cycle in a planar state space.

4.1.8. If we leave the motor running, we may forget start-up transients, and regard the limit cycle itself as the entire state space. Thus, there is only one observed parameter for this driving system: its phase, ϕ. The phase varies from 0 to 2π around the cycle of phases, which is the state space. The frequency of the cycle is supposed to be fixed by the governor. It is not a variable, but a constant in the model.

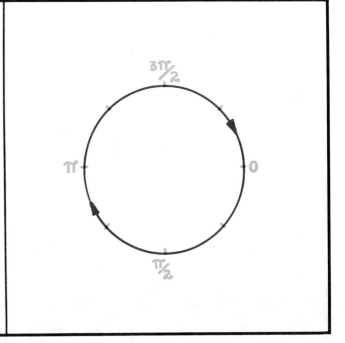

4.1.9. For ease of visualization, we now cut the cycle, and unroll it into a straight line segment. Cut on the right, at *phase zero*. Holding the upper end fast, bend the lower end down, to the left, and up, until straight. Having cut the cycle at phase zero, both ends of this line segment correspond to the beginning of the cycle.

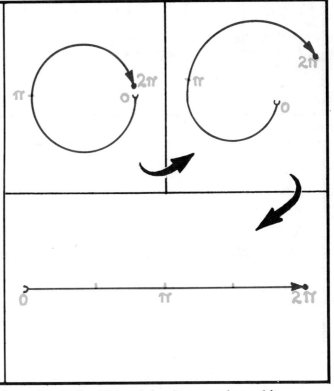

The dot marks *phase zero*, $\phi=0$, the beginning of a cycle. This choice is somewhat arbitrary, but we show here the most common choice, called *the cosine convention*. In this convention, phase zero means the rod has just arrived full left, and is turning to go back. Thus, the pointed top of the lever is *full right at driving phase zero*.

Next, we combine the geometric models for the driven pendulum and the driving motor into a single, combined model. This model represents *the compound system of the two devices, uncoupled.*

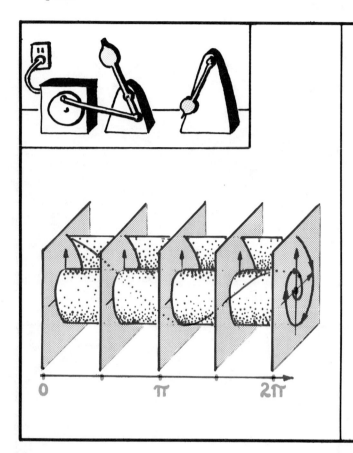

4.1.10. At each point of the unrolled cycle of phases of the driving motor, place an identical copy of the plane of states of the driven system. Orienting these planes vertically as shown, we may think of the scheme as a deck of cards on edge. Each card has the phase portrait of the damped pendulum printed on it. This is an example of the *Cartesian product construction* described in Volume 0. Every point in the resulting three dimensional scheme represents, simultaneously, a state of the pendulum, *(A,R),* and a phase of the driver, ϕ. The three dimensions, thus, represent the observed parameters, *(ϕ,A,R),* of the combined (but uncoupled) system. Comparing with Figure 1.4.7, notice that the *phase* of the driving oscillation has replaced *time* as the parameter. Thus, the driving oscillation has become the clock. During one cycle of this clock, the state of the driven oscillation approaches its attractor, as shown by the exemplary trajectory (red) in this illustration.

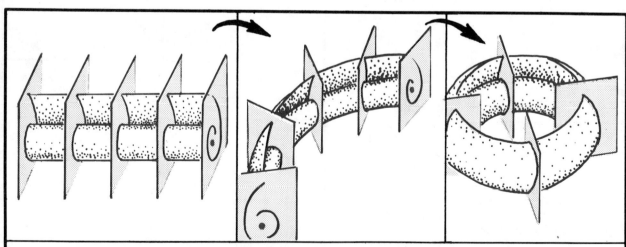

4.1.11. Finally, to get the correct model for the combined system, we roll up the cycle of phases again, carrying the deck of cards along with the cycle. Bend the left end down, to the right, and up again, until the two ends meet at the right. Glue the two end cards together. This *ring model* is the geometric model for the combined system.

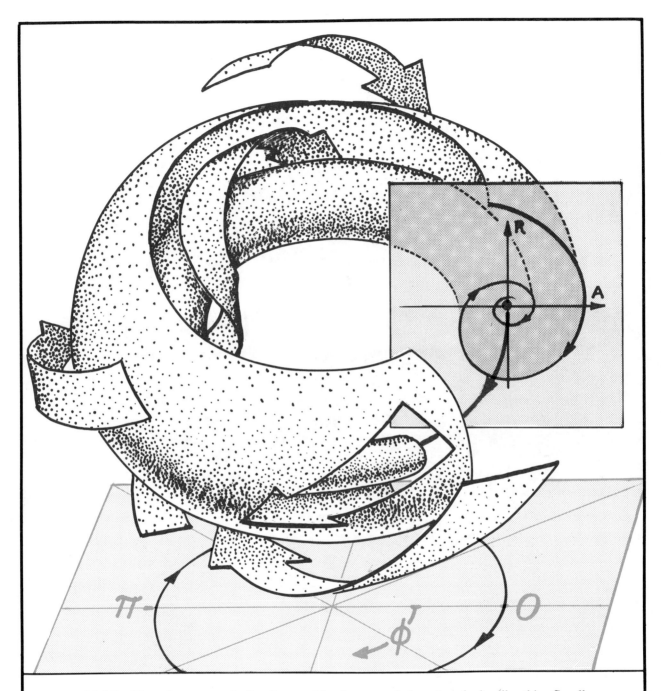

4.1.12. The phase portrait for the combined, uncoupled system looks like this. Scrolls within the ring contain the trajectories, which spiral around the red cycle in the middle of the ring. This is an attractive limit cycle for the combined dynamical system. It represents the pendulum bob coming to rest, as the driving motor keeps on running at its regulated frequency. Each scroll is actually a cylinder, rolled up like a pant leg. Also, it is an *invariant manifold* of the flow. This means, simply, that it is a collection of trajectories. No trajectory enters or leaves a scroll. A slice has been removed from the scroll at phase zero for better visibility.

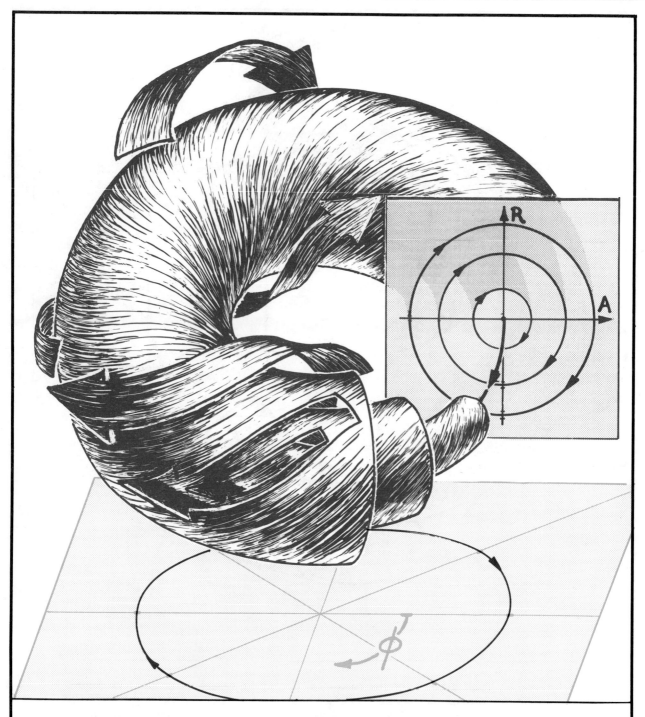

4.1.13. For comparison, here is the phase portrait for the combined dynamical system, in the case of the *undamped* pendulum, with the same driving motor. The scrolls are replaced by concentric *tori*. Each card of the deck is printed with concentric circles. The trajectories of the combined system spiral around these tori, which are invariant manifolds. The central cycle (red) is *not a limit cycle* for nearby trajectories. These concentric tori have also been cut through, at phase zero, for visibility.

The trick of cutting through the combined phase portrait at a fixed phase, for better visualization, came from the early experimentalists. It is now called stroboscopy: Rayleigh names Plateau (1836) as the inventor[1].

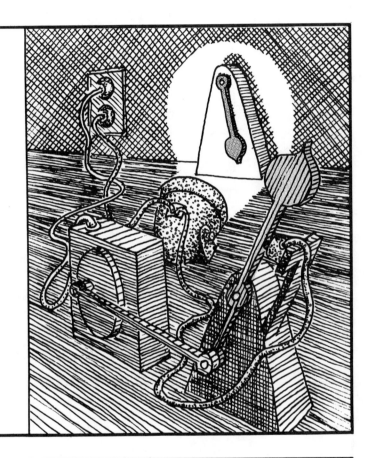

4.1.14. The strobe lamp is aimed at the driven bob. It is turned on momentarily, when the driving (green) pendulum contacts the microswitch, at phase zero. Lord Rayleigh used the rays of the sun, interrupted by an electrical diaphragm.

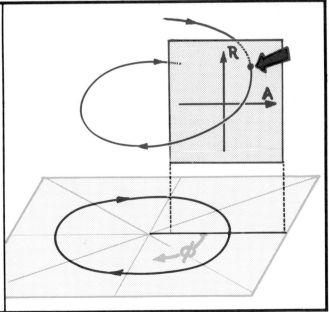

4.1.15. In the light of the stroboscope, the parameters of the driven bob may be observed at the fixed phase of the driving motor. The observed data define a point in the *strobe plane*, the card of the deck corresponding to this fixed phase. At this point, the trajectory of the combined system pierces the strobe plane.

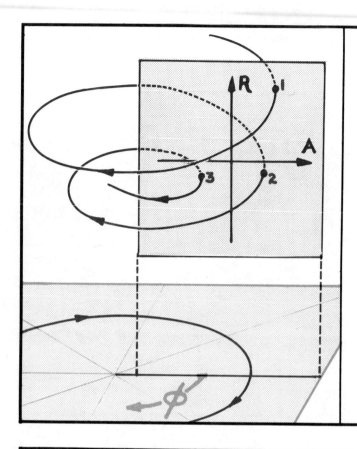

4.1.16. Recording the observations of successive flashes of the strobe light, a sequence of points in the strobe plane is obtained, instead of a continuous trajectory, as the record of motion of the bob. We may call this a *strobed trajectory*.

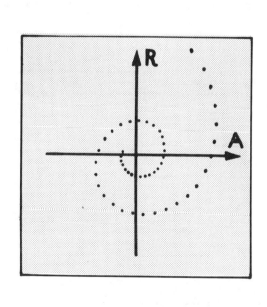

4.1.17. If the driving frequency is relatively fast with respect to the free (unforced) frequency of the driven pendulum, the strobed trajectory will appear to step along a spiral in the strobe plane. In other words, if the driving clock (see 4.1.10) runs quickly, the free pendulum will seem slow. That is, it will run around the spiral almost like a continuous trajectory. Here the driving frequency is approximately 24 times the free frequency.

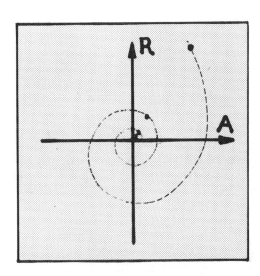

4.1.18. If the driving frequency is about the same as the free frequency of the driven system, the strobed trajectory may appear to walk directly toward the origin, as the pendulum comes to rest. In the strobe light of the experiment, the blue pendulum will seem to swing slowly to rest from one side.

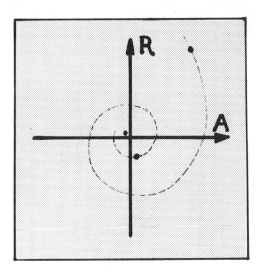

4.1.19. Finally, if the driving frequency is relatively slow with respect to the pendulum, the strobed trajectory will take giant steps, spiraling at a very slow pace toward rest. Here, the driving frequency is approximately 3/4 of the free frequency.

All this regard for the uncoupled system has been for practice in 3D visualization, and getting used to the ring model. In the next section, we will finally connect the two gadgets.

4.2. FORCED LINEAR SPRINGS

At last, let's pin the pendulum to the swinging end of the motor driven lever.

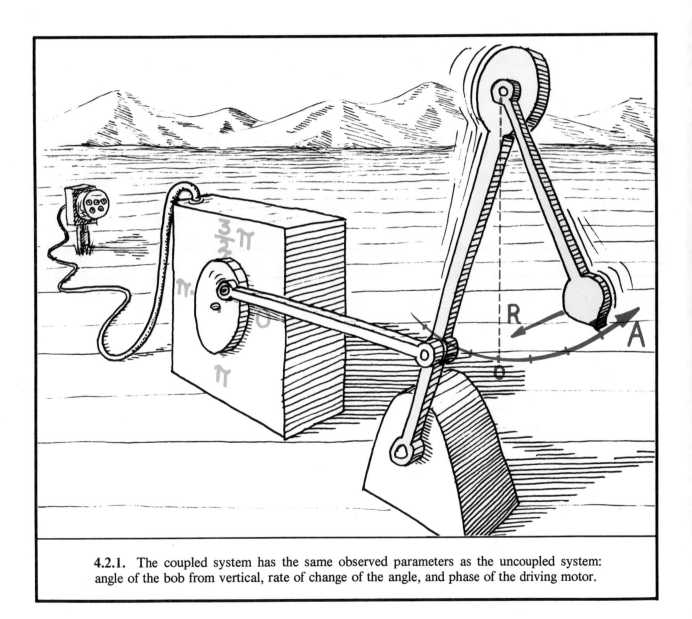

4.2.1. The coupled system has the same observed parameters as the uncoupled system: angle of the bob from vertical, rate of change of the angle, and phase of the driving motor.

The state space is still the ring. The dynamics, however, are changed. The vectorfield is quite different. So are the trajectories, and the phase portrait.

To get the idea of the phase portrait for the coupled system (forced vibrations) a few armchair experiments will be helpful. We call this *the game of bob*.

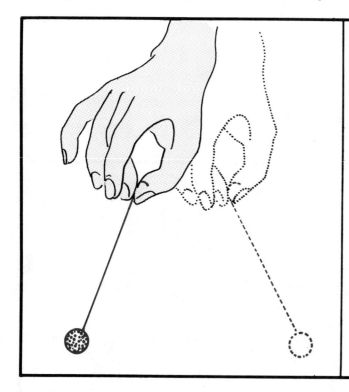

4.2.2. Give your hand a little jerk, then hold it still. The pendulum swings slowly to rest. Notice the frequency. This is the frequency of the free, un-forced pendulum. Now, without chang-ing the length of the string, oscillate your hand back and forth along a hor-izontal line. If the driving frequency of your hand is *slower* than the free fre-quency, the pendulum follows your hand. Instead of coming to rest, it is in *sustained oscillation*.

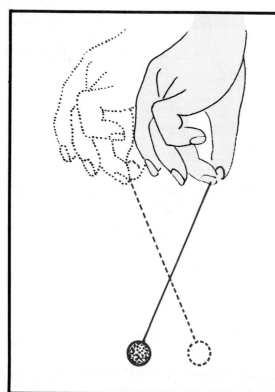

4.2.3.   Start again with your hand and the bob at rest. This time, oscillate your hand *faster* than the free frequency. No-tice that the bob swings *opposite* to the motion of your hand.

If you are in public, that's enough for now. But if nobody is watching, here is another experiment to try.

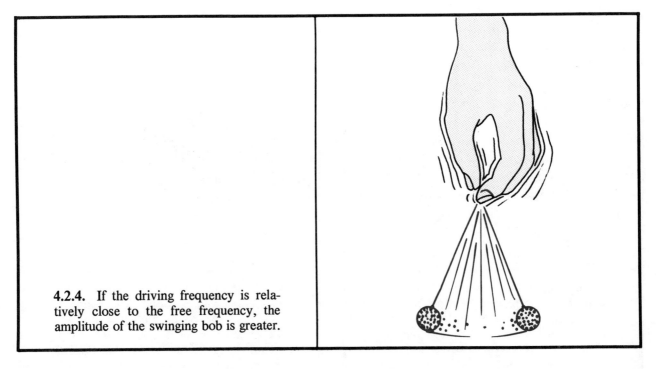

4.2.4. If the driving frequency is relatively close to the free frequency, the amplitude of the swinging bob is greater.

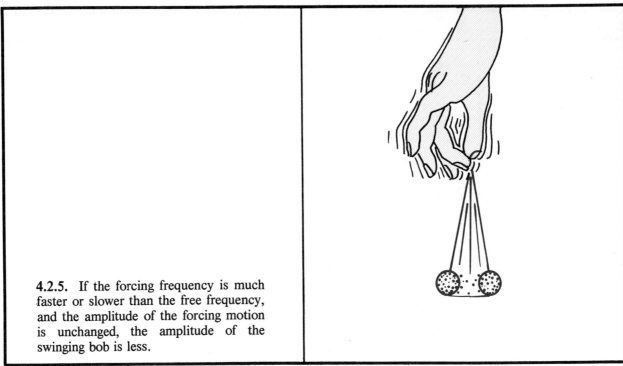

4.2.5. If the forcing frequency is much faster or slower than the free frequency, and the amplitude of the forcing motion is unchanged, the amplitude of the swinging bob is less.

Now let's get on with Duffing's discoveries. First, we will improve the mechanical setup to simplify the analysis.

4.2.6. The driving motor will be the same as in the previous section, but the connecting rod is disconnected from the vertical lever, and attached to a weight. This weight is forced to move horizontally on the tabletop.

4.2.7. As in Chapter 2, we replace the pendulum with a spring. The spring is fixed to a support on the left, and to the weight on the right. The weight is free to move on the tabletop, which resists the motion with friction.

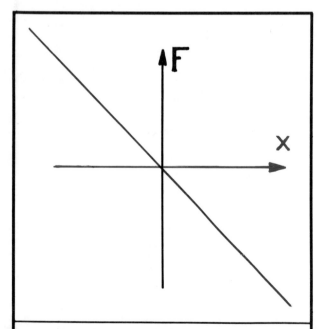

4.2.8. For a start, we assume the spring is linear. This restriction will be removed in the following section.

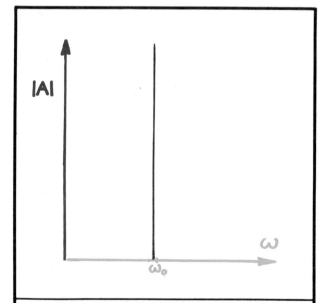

4.2.9. Recall that weights with linear springs have a natural frequency, independent of the amplitude of the oscillation. They make perfect guitar strings.

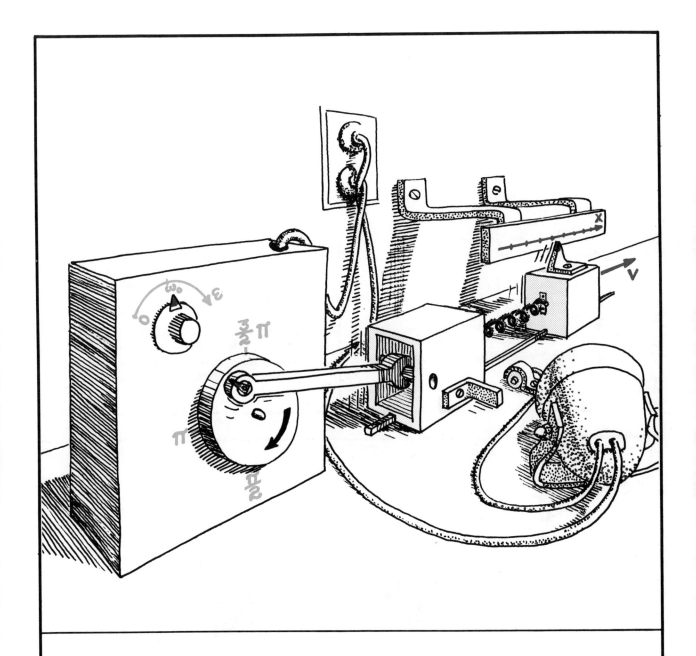

4.2.10. Here is the coupled system. The connecting rod of the driving motor moves the support of the spring, previously fixed to the tabletop, in a horizontal oscillation. For experimental purposes, we include a speed control for the forcing frequency, and a strobe light set for phase zero.

The state space for this forced vibration is again the ring, with observed parameters: phase of the forcing cycle, deflection of the spring, and velocity of the driving weight. As for the phase portrait, we have discovered by experiment that it contains an attractive limit cycle. Thus, in the coupled system, the driven (blue) weight is forced to oscillate by the driving (green) weight.

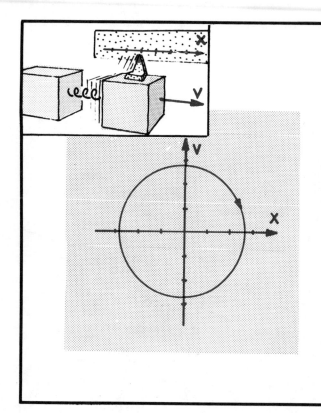

4.2.11. The bob oscillates. To represent this as a limit cycle in the ring, let's begin by recording a full cycle of the bob, in its own, planar, state space. This is *not a trajectory* of a dynamical system in the plane, just a step in a graphical construction. Later we will erase it.

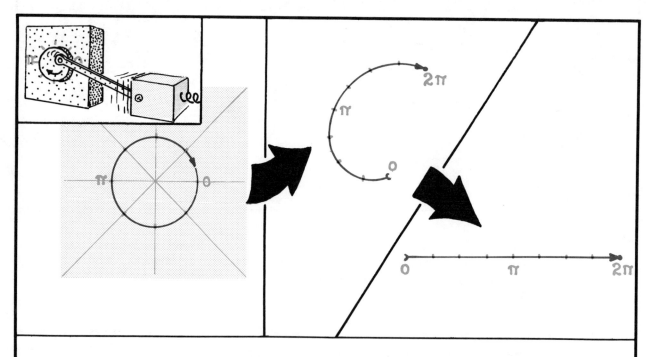

4.2.12. Recall that the state space for the driving motor is a cycle of phases. Earlier, the ring was constructed by cutting this cycle, and straightening it out.

4.2.13. The deck of cards represents the ring, cut and straightened. On each card, we have drawn (in green) the observed cycle of the bob, as shown in the panel before last. This makes a green cylinder in the 3D model.

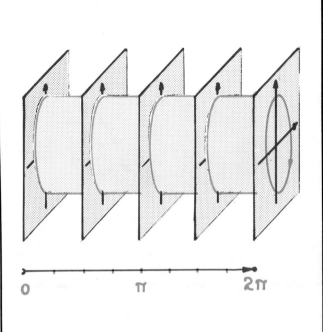

4.2.14. Now, record the observed motion of the combined system in this model, as the bob oscillates, and the driving motor traverses one full cycle. The red curve is this record. It is seen to stay on the green cylinder. It is a trajectory. Note that this illustration closely resembles Figure 1.3.9, with driving phase in place of time, as the parameter.

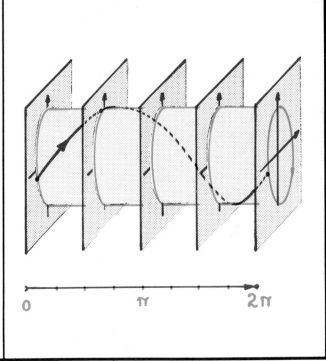

The green cylinder is not a feature of the phase portrait, it is just another step in our graphical construction. Later, we will erase this, also.

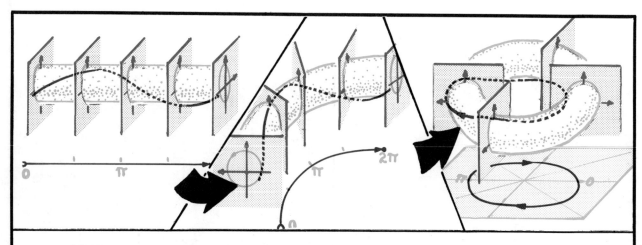

4.2.15. Next step, roll up the deck of cards again, to make the ring. The green cylinder becomes a green torus. The red curve closes up, making a cycle on the green torus.

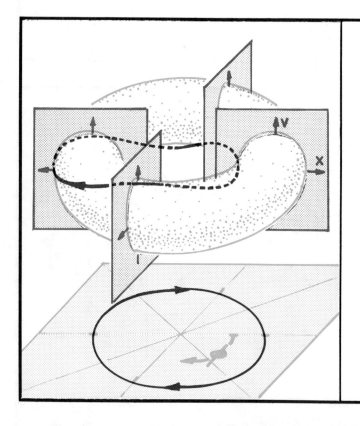

4.2.16. This is the attractive limit cycle known to Rayleigh, Duffing, and all those who have played the game of bob as we instructed. It is called an *isochronous harmonic,* as it completes one full cycle of the bob, to each full cycle of the driving motor.

Now is a good time to erase the green torus. It is not an invariant manifold; most trajectories pass through it. It is just an aid for visualizing the isochronous harmonic, which is the only trajectory on it. Next, let's see what the nearby trajectories actually do.

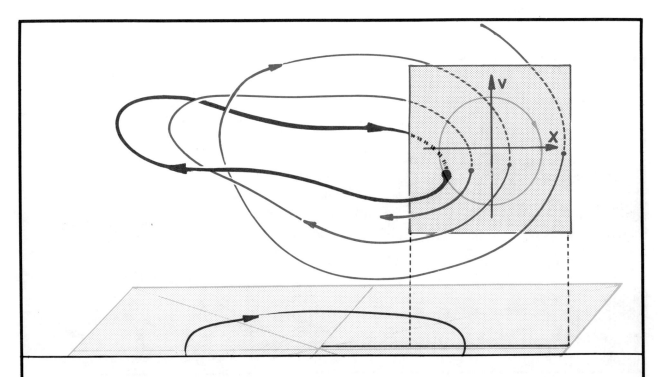

4.2.17. Here we have cut through the ring at phase zero for better seeing. The blue trajectory is a typical one. It spirals around the ring, getting closer to the red attractor with each cycle.

4.2.18. Observing this motion with the strobe light, we plot the tracks of several blue dots in the strobe plane at each flash. This is the strobe trajectory of this motion. The isochronous harmonic meets the strobe plane at a single point, the red one. The blue strobe trajectory approaches closer and closer to the red point. All nearby trajectories approach the red point like this, and may cross the green ring.

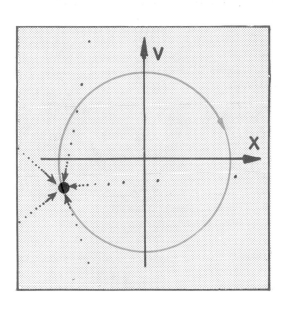

Retracing the steps of Duffing, let's play the game of bob in earnest with this apparatus. Suppose we change the speed of the driving motor.

4.2.19. If the forcing frequency is very slow, the driven (blue) weight will follow the motion of the driving (green) weight. The strobe at phase zero (according to the cosine convention, this means full right, and turning to go back) will flash on the blue weight at its phase zero position.

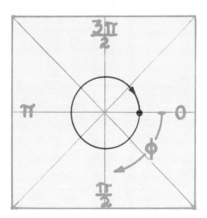 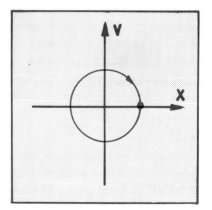

4.2.20. This means that the two separate trajectories are cycles which are *in phase*. That is, the both reach phase zero at the same time. They are both full right, and turning to go back, at this moment.

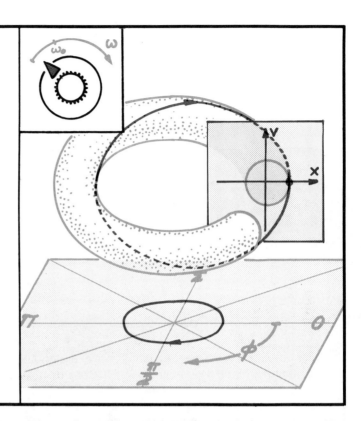

4.2.21. Thus, the combined trajectory, a periodic attractor, passes through the line of constant phase, zero, in the state plane of the driven (blue) weight, (the positive horizontal axis) at the same time the driving phase passes the same value. This occurs just as the strobe flashes, and records the point shown here, in the strobe plane.

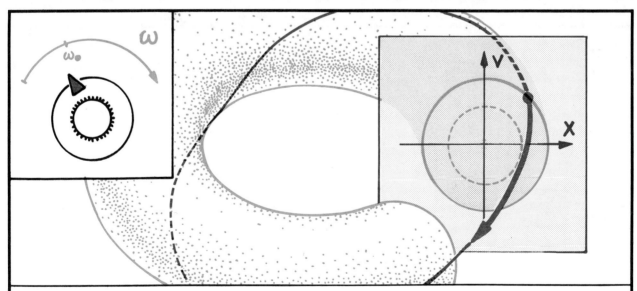

4.2.22. Increasing the driving frequency to a value close to the free frequency of the linear spring and driven (blue) weight, but still smaller, the attractive limit cycle is still an iso-chronous harmonic. But its phase has slipped. Here, the driven (blue) weight is *behind* the driving (green) weight in phase. This is represented by the red strobe point position *above* the horizontal axis in the strobe plane. Further, the amplitude of the driven (blue) weight's motion is larger than before. This is represented by drawing the red limit cycle on a larger imaginary torus.

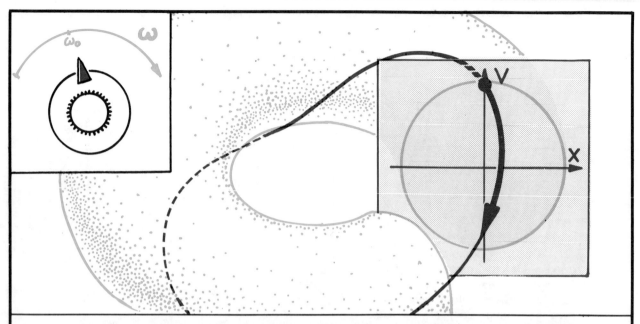

4.2.23. As you may have found in playing the game of bob (4.3.4), the forced oscillation attains its largest amplitude when the driving frequency is equal to the free frequency of the driven oscillator. But the strobe light on our machine reveals further that at this maximum amplitude, the phase of the forced (blue) weight is $\pi/2$ (a quarter cycle) behind the driving (green) phase.

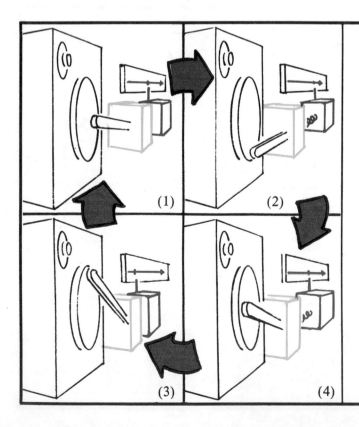

(1)

(2)

(3)

(4)

4.2.24. The mechanics of this phase delay is very well known to children who swing. (4) When the driven weight (yourself) goes through phase π (all the way up in back) the driving force goes through phase $-\pi/2$ (you begin to pump forward). (1) At the bottom of the swing forward, your pump has peaked. (2) At the top of your swing to the front, you begin to pump back. (3) At the bottom of your back swing, your back pump has peaked. (Be careful with the larger amplitudes.)

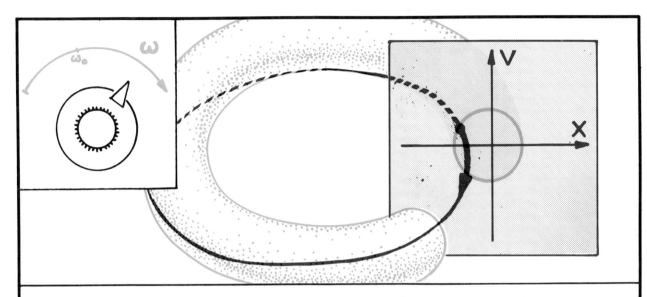

4.2.25. With the driving frequency much faster than the free frequency, the phase of the driven (blue) weight lags a half cycle (-π) behind the forcing (green) weight. Also, the amplitude of the forced oscillation is smaller than it was, when the frequencies were the same. But we know this already, from the game of bob. The smaller amplitude is shown, in this illustration, by the smaller "waist" of the locating torus. Compare with 4.2.23.

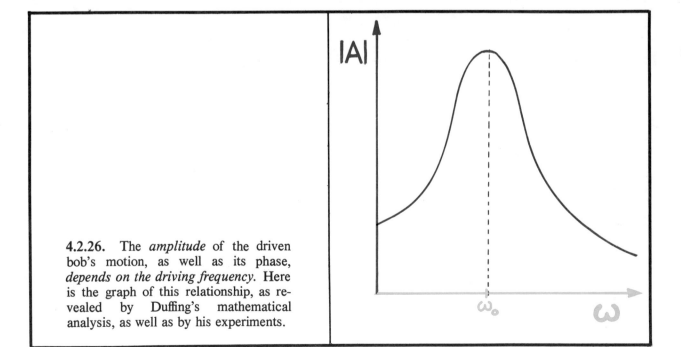

4.2.26. The *amplitude* of the driven bob's motion, as well as its phase, *depends on the driving frequency.* Here is the graph of this relationship, as revealed by Duffing's mathematical analysis, as well as by his experiments.

There are precious few linear springs around, so we really should do this all over again for hard and soft springs. And so, on to the next section for hard springs. Soft springs will be left to the reader as a test.

4.3. FORCED HARD SPRINGS

A hard spring is slightly more realistic than a linear spring. Recall that with a linear spring, if the force required to stretch it a distance x is $F(x)$, then the force required to stretch it twice as far is twice as great, or $F(2x) = 2F(x)$.

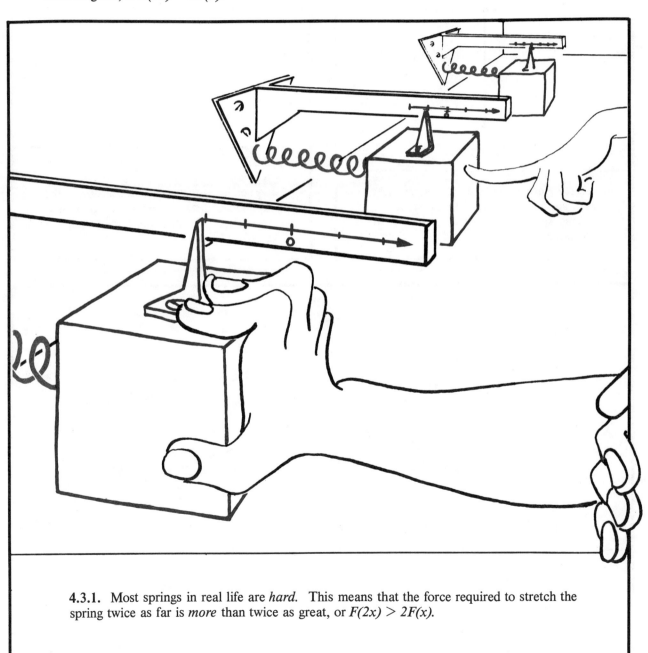

4.3.1. Most springs in real life are *hard*. This means that the force required to stretch the spring twice as far is *more* than twice as great, or $F(2x) > 2F(x)$.

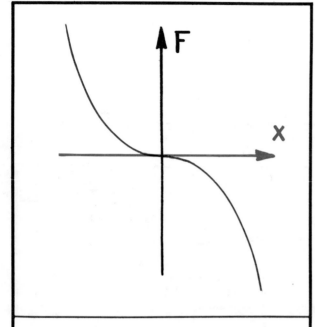

4.3.2. Here is an example of an ideal hard spring. The characteristic function of this spring is a cubic, as shown in this graph.

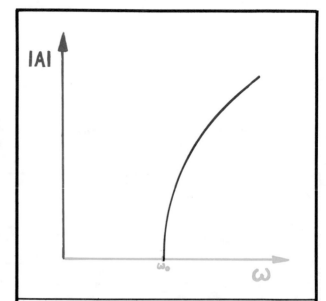

4.3.3. The dynamic consequence of this stiffness is this bend in the response curve: as the amplitude decreases, the frequency decreases—another poor guitar string. The next illustrations explain this bend.

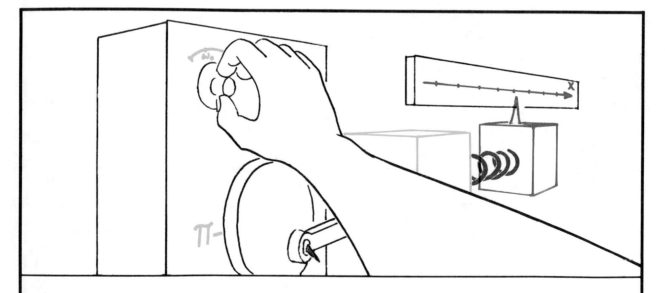

4.3.4. Here is our laboratory apparatus, left over from the experiments of the preceding section. Let's replace the linear spring with a hard one, and observe a sequence of forced vibrations, with different forcing frequencies. The amplitude of the forcing oscillation is not changed in this sequence.

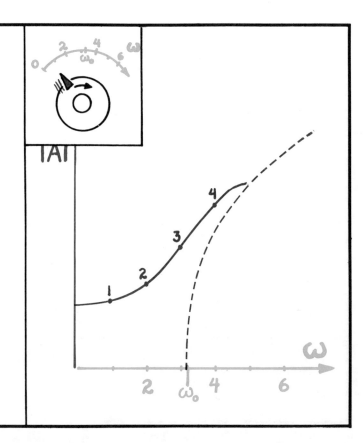

4.3.5. Starting with the speed control at position 1 (slow forcing) and then increasing it, we see a shift in phase, and an increase in amplitude, of the resulting oscillation of the driven (blue) weight. At positions 1 and 2, this is about the same as the linear case. At positions 3 and 4, the effect of stiffness can be observed. The amplitude keeps on increasing, even after the forcing frequency exceeds the free frequency.

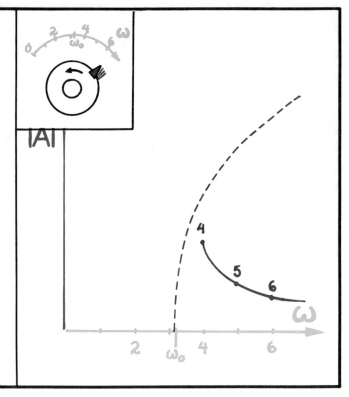

4.3.6. Now start a new sequence of experiments, with the driving motor running at top speed, and decreasing it for each test. Again, the response is much like the linear spring at positions 6 and 5. But at 4, the amplitude of the response vibration suddenly jumps to the higher value. This higher amplitude is the same one found in the first sequence of observations, after increasing the speed from position 3 to 4.

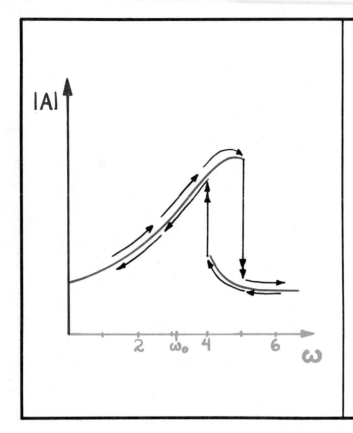

4.3.7. Moving the speed lever all the way to slowest, then back to fastest, back to slowest, and so on, we observe this *hysteresis loop of Duffing.* At 4 and decreasing, the amplitude suddenly increases. At 5 and increasing, it suddenly decreases. If you don't believe it, try in out. Because a pendulum behaves like a soft spring, you may observe this phenomenon in the game of bob.

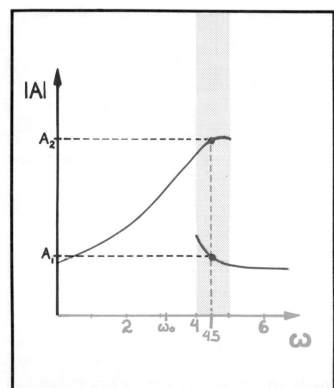

4.3.8. If you do believe it, then you agree that when the speed control of the driving motor is between positions 4 and 5, *there are two periodic attractors* in the phase portrait of the forced system.

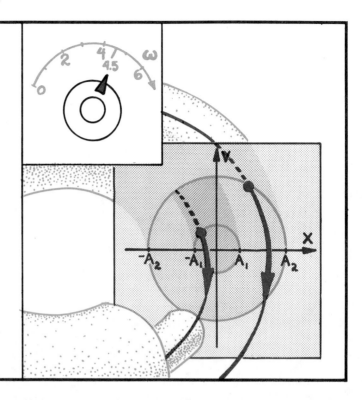

4.3.9. One of the attractive limit cycles, corresponding to the smaller vibration, is close to a half cycle out of phase with the driver. (See 4.2.25.) It circles around a smaller (imaginary) torus. The other, corresponding to the larger vibration, is roughly in phase with the driver. (See 4.2.22.) It circles around a larger (imaginary) torus.

Both of these limit cycles are isochronous harmonics. Neither of the tori is an invariant manifold —most trajectories go through them. Some nearby trajectories tend asymptotically to one of these attractors, some to the other. There are two basins, as in the examples of Sections 1.6 and 2.2.

The two basins are divided by a separatrix. Where is it located, in the phase portrait?

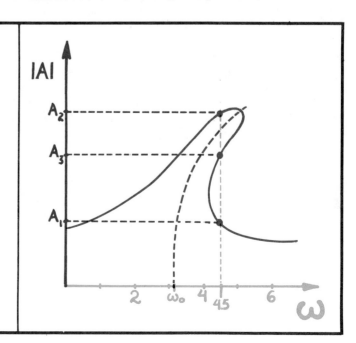

4.3.10. In the response diagram, the two branches of the response curve, observed in the experiments described above, are *connected* as a smooth curve. The connecting segment corresponds to another limit cycle in the phase portrait—one which is *experimentally invisible*.

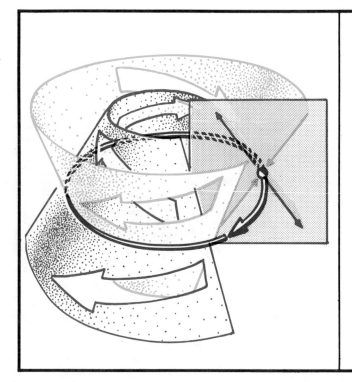

4.3.11. The invisible third limit cycle is a *saddle cycle.* These occur in state spaces of dimension three or more. A saddle cycle is the periodic analogue of a saddle point. That is, they are related to periodic attractors as saddle points are related to point attractors. In three dimensions, the *inset* of a saddle cycle is a surface, a deformed cylinder belted by the saddle cycle. The inset is shown in green in this illustration. It is a *separatrix,* it separates two basins. The *outset* of a saddle cycle is another deformed cylinder, which crosses through the inset where it is belted by the saddle cycle. The outset is shown in blue. Half of the outset is in one basin, half in the other. These halves are divided by the cycle itself, shown in red.

This green inset is an *invariant manifold.* It consists of trajectories which stay on it forever, and tend asymptotically toward the red saddle cycle as time increases. The saddle cycle is the omega-limit set for the trajectories on the green inset, as explained in section 1.5.

To better visualize the phase portrait, we may use the trick immortalized by Poincaré: extract the strobe plane.

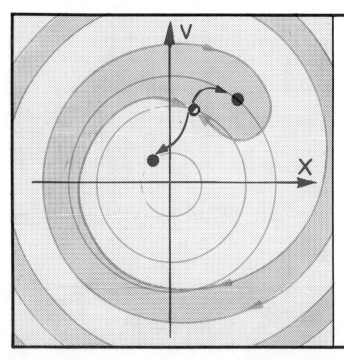

4.3.12. Here the strobe positions of the three limit cycles are shown as red points in the strobe plane. The solid red dots represent attractors. The small half-filled circle denotes the saddle cycle. The inset of the saddle, two green spirals, intersect the strobe plane in two curves, shown here in green. They divide the strobe plane into two regions: the darkly-shaded tear drop shape, and the rest. These two regions comprise the *basins in cross-section,* that is, in the strobe plane. Note that the saddle point in the cross-section is *between* the two attractors, both in phase and in amplitude.

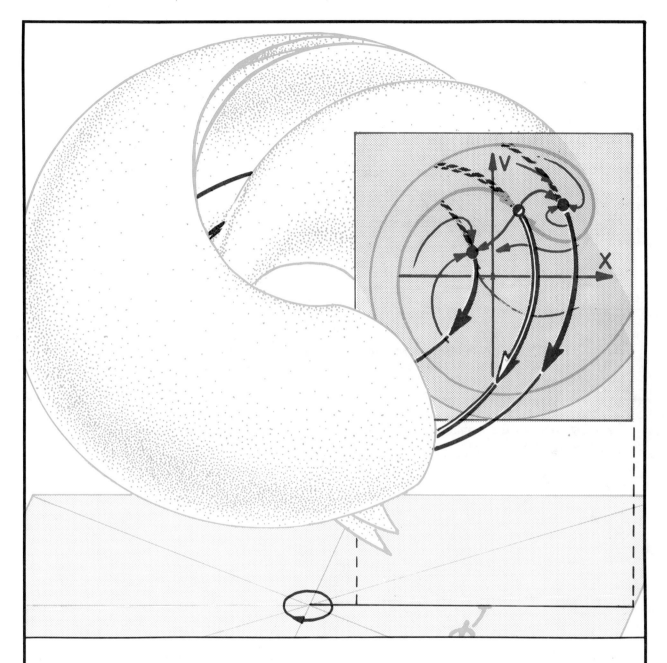

4.3.13. To visualize the basins in full, as three-dimensional regions, we repeat the strobe process at many successive phases of the driving motor. The full basins, revealed by this technique, look like this: yin-yang in 3D. The tear drop revolves once as the section progresses around the driving cycle. The three red cycles are the isochronous harmonics discovered by Duffing.

All this from experimenting with different forcing frequencies, at the same forcing amplitude. What if we changed the amplitude?

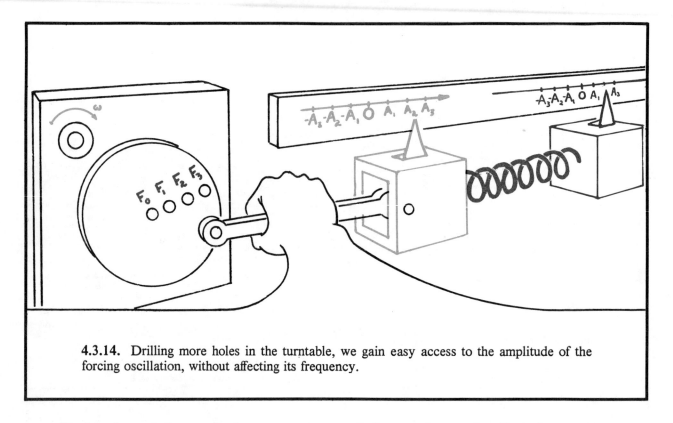

4.3.14. Drilling more holes in the turntable, we gain easy access to the amplitude of the forcing oscillation, without affecting its frequency.

Having changed the amplitude, we may repeat all the experiments described above, with the frequency changed variously.

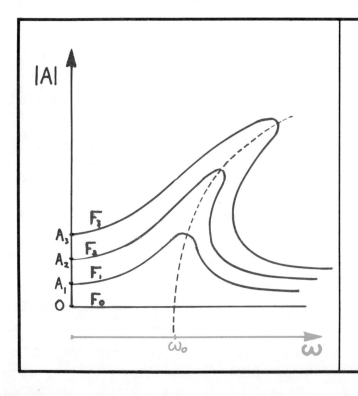

4.3.15. The results are always similar, except that the location of the response curve is shifted, kink and all. Here are the response curves for three different amplitudes. The larger the amplitude, the higher the response curve in this diagram. Notice that the lower curve (smallest forcing amplitude) looks rather like the response curve of a linear spring.

4.3.16. Regarding the response as a function of the two control parameters, frequency and amplitude of the forcing oscillation, we find that the curves on the right are successive slices of a kinked surface in a three-dimensional response diagram.

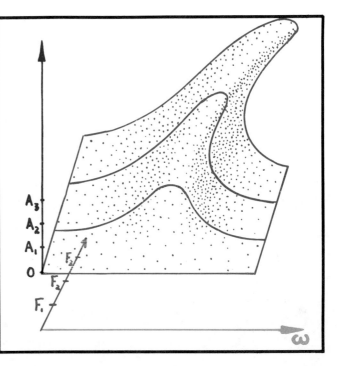

4.3.17. If you have already heard about catastrophe theory, you may recognize this as the *cusp catastrophe[3]*. Here, we have added the *hysteresis loop* to the cusp figure.

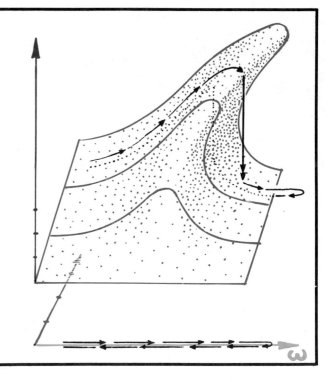

Actually, since Duffing, there has appeared an enormous literature on the forced pendulum. Much of this is devoted to additional attractors which have been found in Duffing's phase portrait, which are *non-isochronous harmonics*.

In the next section, we give a brief introduction to these other harmonics.

4.4. HARMONICS

When the driving amplitude is fixed but the frequency is changed in a sequence of experiments, as described in the preceding section, unusual motions of the bob turn up when the driving frequency is twice the free frequency, or half of it, and so on. These are the *non-isochronous harmonics*. Here is the simplest example, sometimes called the *second harmonic*. Suppose that the driving motor is turning at about half of the free frequency of the spring-bob system. We find that the driven system responds with a sustained oscillation at about the free frequency.

Get out your game of bob again for this one.

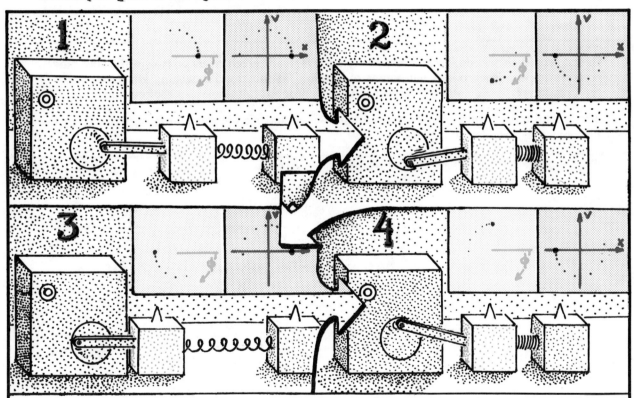

4.4.1.   Here are four strobe views of the second harmonic motion.
*1. Driver phase 0, driven phase 0. They are both at the beginning of a cycle, in phase.
*2. Driver phase $\pi/2$, driven phase π.
*3. Driver phase π, driven phase 0. Since being in phase in 1, the driven (blue) weight has completed a full cycle, while the driving turntable has completed only half a turn.
*4. Driver phase $-\pi/2$, driven phase π.
*1. Driver phase 0, driven phase 0. Since 1 above, the turntable has completed a full turn at last, while the driven weight has done two swings.

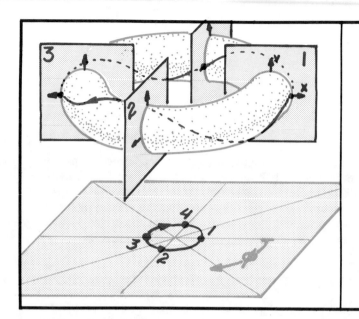

4.4.2. Let's now pick up the four cards (strobe planes) of the preceding panel, and place them in their proper places within the ring model. Interpolating the rest of the trajectory, we obtain this red trajectory. It is an attractive limit cycle, which winds twice around the waist of the imaginary green torus, while making one full circuit of the ring. It represents the second harmonic of the driven spring-bob.

In the preceding section, there was already an imaginary green torus or two in the ring model, representing the isochronous harmonics of Duffing. If this double twisting attractor is also in the ring, how is it situated with respect to the isochronous harmonics, or the rotating tear-drop?

The isochronous harmonics, the second harmonics, and all the other little harmonics, all live harmoniously together, in the ring model.

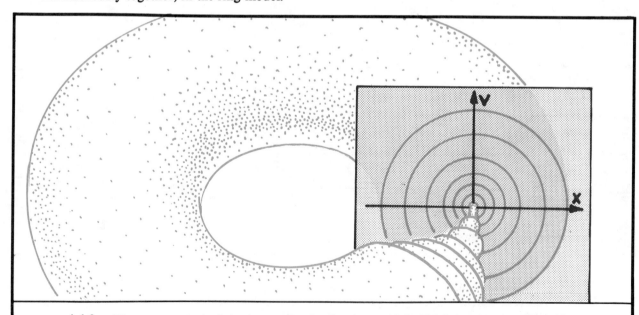

4.4.3. Here are several of the harmonics in the ring model. Each has its own home, in one of the concentric rings. When the forced oscillation creates a harmonic (periodic attractor) in the compound system with a non-isochronous frequency, *the amplitude of the harmonic is smaller.* Thus, it winds around a locating torus with a smaller waist. Thus, the outer rings in this illustration contain the isochronous harmonics. The smaller rings house the more exotic harmonics, just as in the Inferno.

To get a clear picture of the harmonic vibrations of a forced pendulum or spring, we need a certain familiarity with trajectories on a torus.

We pause now for a short course in toral arrangement.

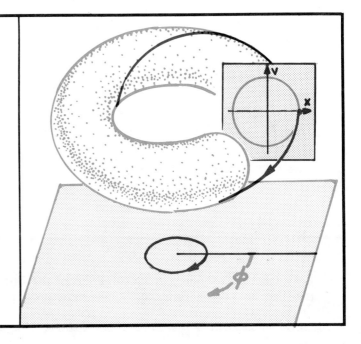

4.4.4. Earlier, in 4.3.21, we introduced a phantom torus to depict the isochronous harmonic, in phase with the driving oscillation.

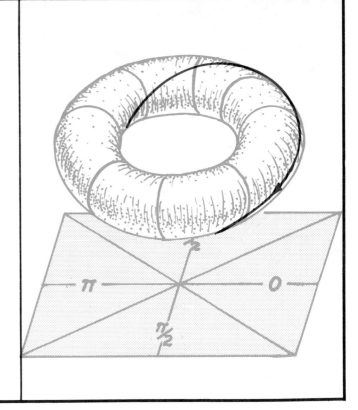

4.4.5. We may regard the state space as a cyclical deck of cards. Each card slices the phantom torus in a vertical (green) circle. thus, the torus is made of green circles. Each of these corresponds to a particular phase of the driving oscillation of the green weight.

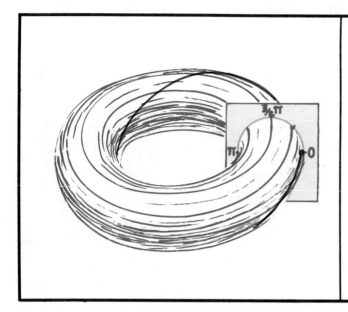

4.4.6. On the other hand, we may think of the torus as being made of these horizontal, blue stripes. Each of these horizontal circles corresponds to a particular phase of the driven (blue) weight.

Let's take an imaginary green torus out of the ring. Slice vertically through it at the strobe plane on the extreme right, corresponding to green phase zero.

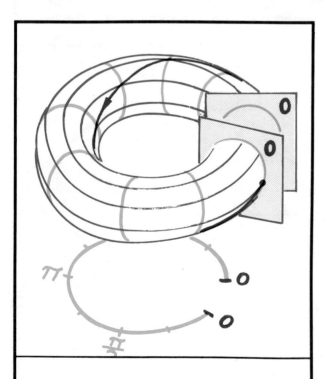

4.4.7. Grab the loose end closer to us. Pull it front, right, and push it back, to straighten it out.

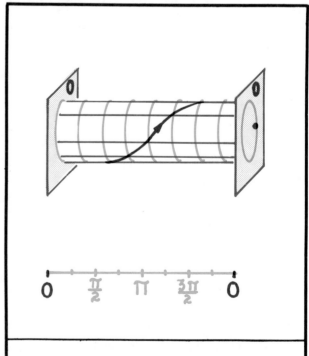

4.4.8. Straightening out the tube and reducing the scale of the green phase, we have a section of a cylinder. Both ends correspond to green phase zero.

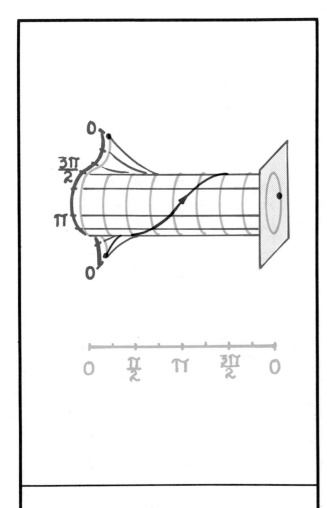

4.4.9. Now cut the back of the cylinder, along the line of zero phase of the blue cycle. Push the lower edge down, while pulling the upper edge up.

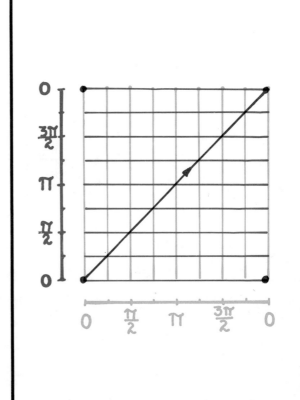

4.4.10. We obtain this flat rectangle. Both horizontal edges correspond to blue (driven) phase zero while both vertical edges correspond to green (driver) phase zero.

The red trajectory turns out to be a straight line through two corners, in the case of a linear spring.

This red trajectory represents an isochronous harmonic, with both cycles in phase. Thus the driving (green) weight and the driven (blue) weight begin together, each at phase zero. After one cycle of each, they return to phase zero simultaneously. To be sure of this, move along the red trajectory, little by little. At each point, observe the turntable phase (by looking down to the lower edge) and the bob phase (by looking left, to the vertical edge) and see how they are changing.

Here is another translation exercise, from toral to flat representation.

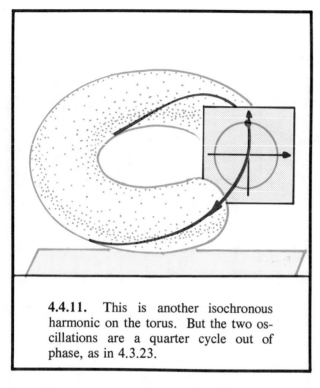

4.4.11. This is another isochronous harmonic on the torus. But the two oscillations are a quarter cycle out of phase, as in 4.3.23.

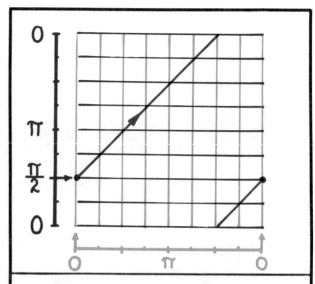

4.4.12. The phase relationship is easier to analyze in the flat representation. On the left edge, driving weight is at phase zero, while driven weight is at phase $\pi/2$.

To step along this red trajectory, you must remember that when you fall off the top of the rectangle, you reappear at the bottom. Likewise, when you run off the right edge, you reappear on the left. It's just like a TV screen.

That finishes the digression on toral arrangement. Now, back to harmonics. The ideas developed, originally, in the context of musical instruments.

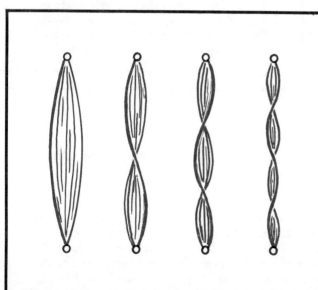

4.4.13. A guitar string vibrates in different *modes*. Explained mathematically by Euler, they were exhaustively studied in experiments by Pythagoras. The sound of each of these modes is a pure tone. The mode shown on the left is called the *fundamental*. The frequency of each mode is an integer multiple of the fundamental frequency.

Musical harmonics may be plotted on a torus. But instead, we will return to the context of the driving and the driven weights. In this context, the driving weight plays the role of the fundamental tone. We shall explain, in sequence, *ultraharmonics, subharmonics,* and *ultrasubharmonics.*

This is an example of an ultraharmonic, in toral and TV representations.

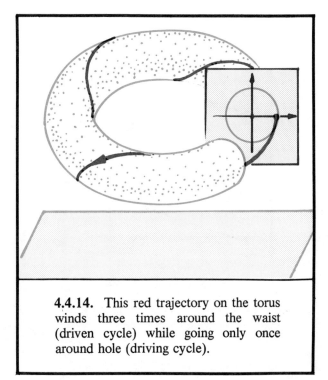

4.4.14. This red trajectory on the torus winds three times around the waist (driven cycle) while going only once around hole (driving cycle).

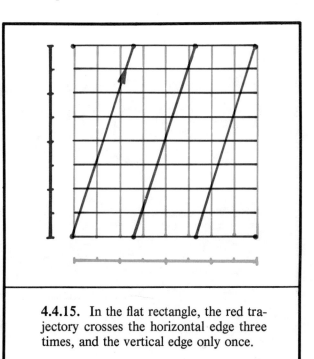

4.4.15. In the flat rectangle, the red trajectory crosses the horizontal edge three times, and the vertical edge only once.

This example is an ultraharmonic of *harmonic ratio 3:1.* That is, 3 bob cycles occur during 1 turntable cycle. There are ultraharmonics of all integer ratios.

4.4.16. Here is the general ultraharmonic. There are an arbitrary number of driven cycles, say *P,* for one turntable cycle. The harmonic ratio is *P:1.*

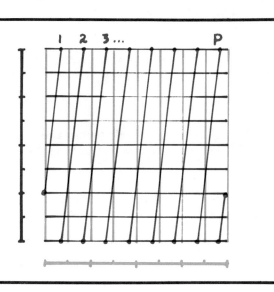

A subharmonic is characterized by slower motions of the driven weight. The driver must complete several cycles, before the driven weight completes one.

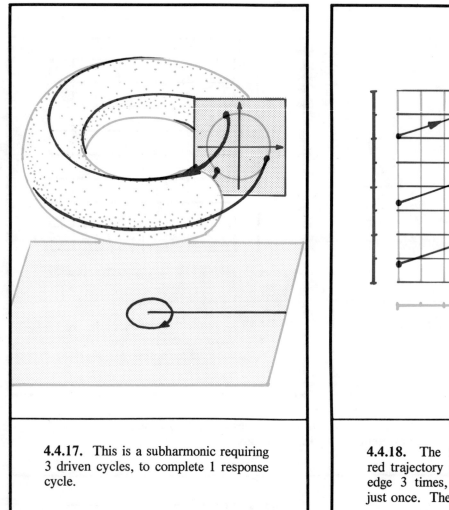

4.4.17. This is a subharmonic requiring 3 driven cycles, to complete 1 response cycle.

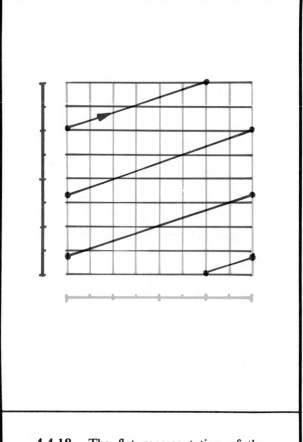

4.4.18. The flat representation of the red trajectory wraps around the vertical edge 3 times, and the horizontal edge just once. The harmonic ratio is 1:3

The general subharmonic requires some number of turntable cycles, say Q, for one cycle of the bob. The harmonic ratio in this case is *1:Q*.

Isochronous harmonics have harmonic ratio 1:1. Isochronous, ultra, and subharmonics are special types of the general harmonic, also known as the *ultra-subharmonic. These have harmonic ratios P:Q, with any integers for P and Q.*

The next example gives rise to the musical interval do-sol (a fifth). The harmonic ratio is 3:2.

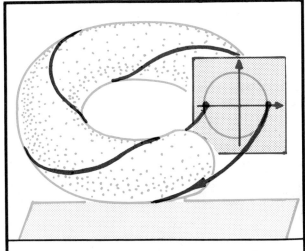

4.4.19. The red trajectory goes twice around the hole, each time making a turn and a half around the waist.

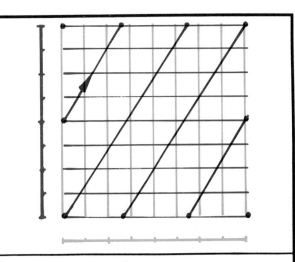

4.4.20. In the flat representation, it transits the horizontal 3 times, and the vertical edge twice.

You see now how to draw a general harmonic, with any rational number *P:Q* as ratio. Beware: we have drawn the trajectories on the flat rectangle more or less straight, for simplicity. But in actuality, they may be very curved.

How about a harmonic with an irrational ratio?

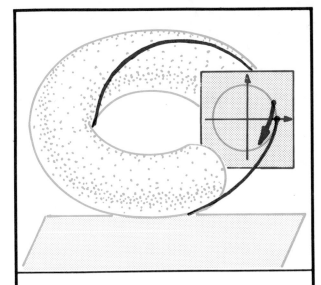

4.4.21. Here is a piece of a red trajectory. If we drew it all, the torus would be almost all red. It goes round indefinitely, without closing. This is the *solenoid* of Section 1.4.

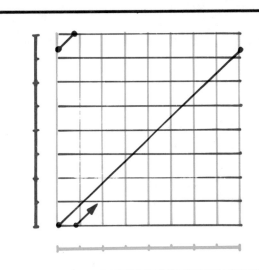

4.4.22. Again we have chosen the simplest situation to draw: the trajectory becomes a straight line in the flat rectangle. But after each wrap-around, it hits a blank space upon re-entry. It never closes.

As we shall see in Volume 2, this situation is very rare. Even in the case of actual harmonics (rational harmonic ratios) the more exotic ratios are relatively rare, as experimentalists have found[4]. They are restricted to very small rings in the ring model. This means they occur with very small amplitudes.

There is much more to be learned from Duffing's game of bob, as we shall see in Volume 2. But now let's move on to the penultimate topic in classical dynamics —compound oscillations.

5.   COMPOUND OSCILLATIONS:
     INVARIANT TORI IN 3D FROM HUYGHENS TO HAYASHI

The considerations of Rayleigh, on the question of the production of sound by bowed strings in particular, evolved into a large branch of dynamics, called *forced oscillations,* or *non-linear vibrations*. The first results were obtained by Duffing, in 1918, for the forced pendulum, has already been described in the preceding chapter. In the following decade, similar results were obtained by Appleton, Van der Pol, Andronov, and others for the periodic, forced perturbation of a self-sustained oscillator. In this chapter, we describe their results. These discoveries for forced oscillators are subtly different from Duffing's observations of forced pendula, described in the preceding chapter. There, we forced a point attractor. Here, we will force a periodic attractor.

5.1. THE TORUS MODEL FOR TWO OSCILLATORS

The first step toward understanding forced oscillations is to make a geometric model for the states. We will use the torus, now familiar from the previous chapter. Then, we describe the dynamics on the torus corresponding to two *uncoupled* oscillators. In the following section, we allow coupling between the two oscillators, describing the dynamics on the torus. In the third section, we introduce the *ring model* for forced oscillations.

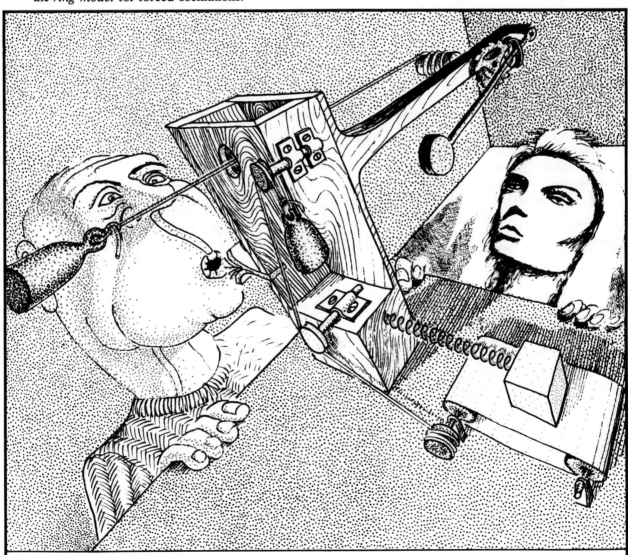

5.1.1. These mechanical systems have dynamical models, described in Chapter 3, with similar phase portraits. They are self-sustained *oscillators*.

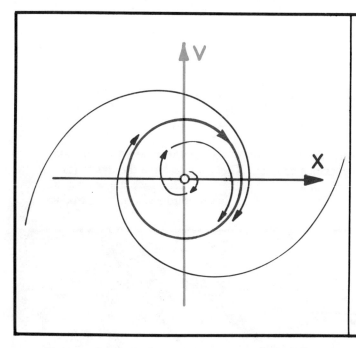

5.1.2. Their phase portraits contain attractive limit cycles, that is, periodic attractors. After the start-up transient dies away, the trajectory follows the attractive limit cycle.

For the sake of visualizing the asymptotic behavior of the oscillators, we may ignore the transient behavior in the dynamical model. Thus, the two dimensional model (the plane with a limit cycle around the origin) may be replaced by the limit cycle itself, standing alone.

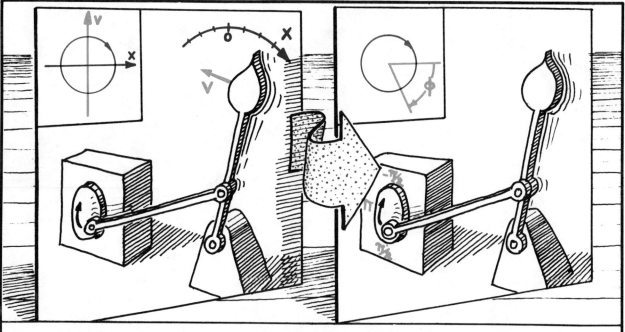

5.1.3. Recall that in the last Chapter, we introduced the *reduced model* for the states of an oscillator. The state of the oscillator is represented by a point of this cycle, corresponding to its *phase*. The two-dimensional model for the states of one oscillator is thus replaced by a one-dimensional model. We allow the plane to fade away, while the cycle remains.

Next, we consider two different oscillators. If the two oscillators are physically separate, the motion of each is uninfluenced by the other. We say they are *uncoupled oscillators*.

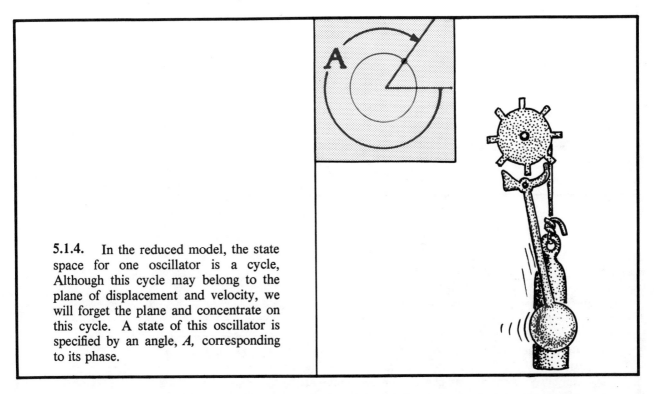

5.1.4. In the reduced model, the state space for one oscillator is a cycle, Although this cycle may belong to the plane of displacement and velocity, we will forget the plane and concentrate on this cycle. A state of this oscillator is specified by an angle, *A*, corresponding to its phase.

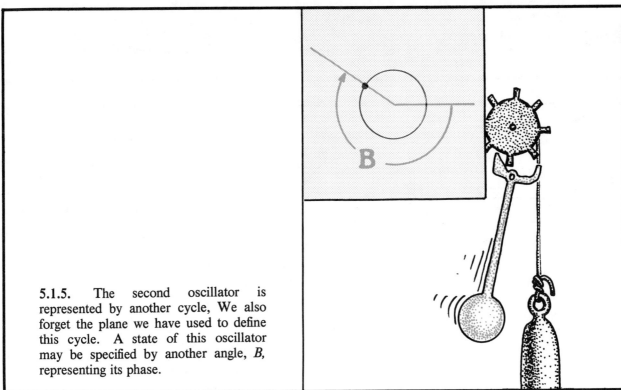

5.1.5. The second oscillator is represented by another cycle, We also forget the plane we have used to define this cycle. A state of this oscillator may be specified by another angle, *B*, representing its phase.

The two oscillators may be described simultaneously in a single state space as follows.

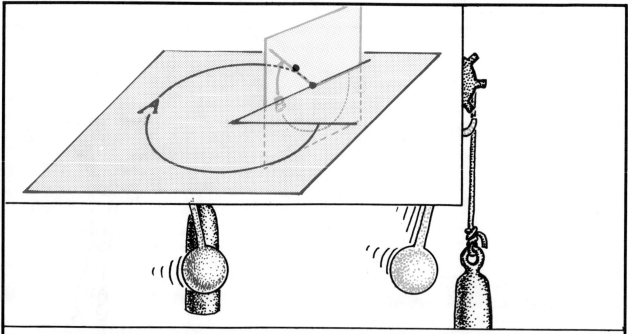

5.1.6. First, recline the planar model for the blue oscillator horizontally. As we saw above, each point of the limit cycle in this plane, *A*, describes a unique phase of the blue oscillator. At this point, *A*, we erect a planar model of the green oscillator standing vertically. We imagine this plane to be *perpendicular* to the limit cycle of the blue oscillator. Within this vertical plane, we visualize the limit cycle of the green oscillator. Each point of this limit cycle, such as *B*, represents a unique phase of the green oscillator. The red point in this drawing is described exactly by the *two phases*, *A* and *B*. This pair of phases, *(A,B)*, describes the red point, and represents the state of the *combined system* consisting of the two oscillators.

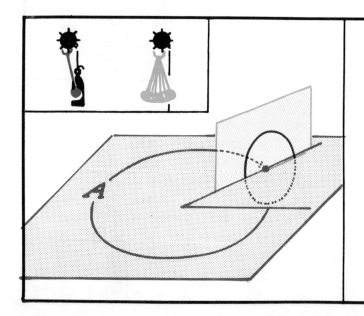

5.1.7. If the blue oscillator is stuck at phase *A* and the green oscillator moves through a full cycle, the red point describing the combined system traverses this red cycle.

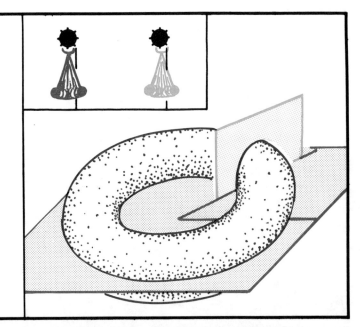

5.1.8. If the blue oscillator moves through a full cycle, the red cycle is pushed around the horizontal cycle, sweeping out this red torus.

Reducing both oscillator models from planes to cycles, the planes in the preceding panels fade away, leaving just the torus. The state space for the combined system of two oscillators is this torus, which is the Cartesian product of the two cycles[2]. The full model for two oscillators is four-dimensional, but this doubly reduced model has only two dimensions. It is easier to visualize.

This all seems very much like the imaginary little green tori in the ring model for the forced pendulum of Duffing, constructed in the preceding chapter. Yet we took pains to point repeatedly out that the green tori were not *invariant manifolds*. Trajectories went right through them as if they were not even there. The red torus is different. It is the state space, there is a dynamical system on it modeling the two oscillators, and the trajectories must stay within the torus.

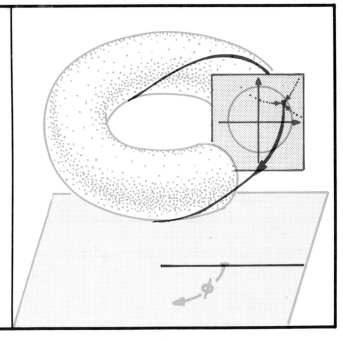

5.1.9. Here is a reminder about the green torus in the ring model for the forced pendulum, in the preceding chapter. It served, in a graphical construction, to locate an attractive limit cycle. Trajectories go right through it. We cannot reduce the ring model for the forced pendulum of the preceding chapter to the green torus.

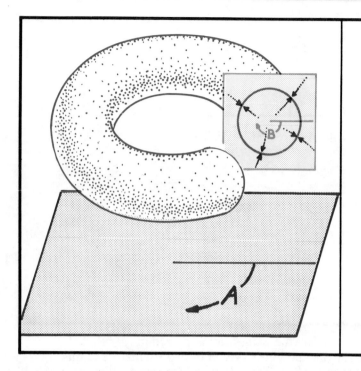

5.1.10. Here, for contrast, is the red torus. This is an attractive, invariant manifold for a dynamical system in another ring model. It is made of trajectories which stay within it. We are regarding this as the state space, ignoring the rest of the ring.

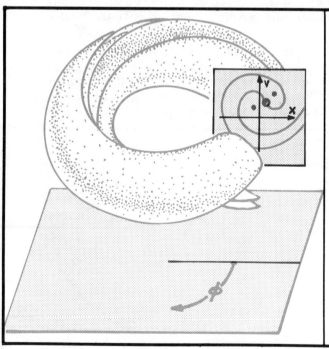

5.1.11. Although the red torus is not an analogue of the green tori in the ring model, it does have a relative in the forced pendulum context. The separatrix dividing the basins of the two isochronous harmonics found by Duffing is also an invariant set.

Having constructed the state space for the two oscillators, we may now describe the dynamical system for compound oscillations, by *coupling* the two oscillators.

5.2. THE TORUS MODEL FOR COUPLED OSCILLATORS

Any two dynamical systems may be combined into a single system by the Cartesian product construction. A small perturbation of this combined system is called a *coupling* of the two systems. For example, the Cartesian product of two one-dimensional systems is a two-dimensional system, as described in the preceding section. In this section, we introduce a perturbation, or *coupling,* in the model for two oscillators. We will obtain a dynamical system on the torus as a geometric model for the behavior of the two coupled oscillators.

5.2.1.   An example is provided by coupling two clocks. This particular system was observed by Christiaan Huyghens, an outstanding dynamicist of the 17th century. He noticed that two clocks hung on the same wall tend to synchronize, and suspected this *entrainment* phenomenon was caused by nonlinear coupling through the elasticity of the wall. The full explanation of entrainment is a recent result in dynamical systems theory. Due to Peixoto, it is described in Volume 2 of this series.

5.2.2. In many applications, two parameters are observed, and both are periodic. The torus will provide a geometric model for all of these empirical situations.

Let's choose a particular situation to model: the two clocks observed by Huyghens. To begin with, consider *two uncoupled clocks*.

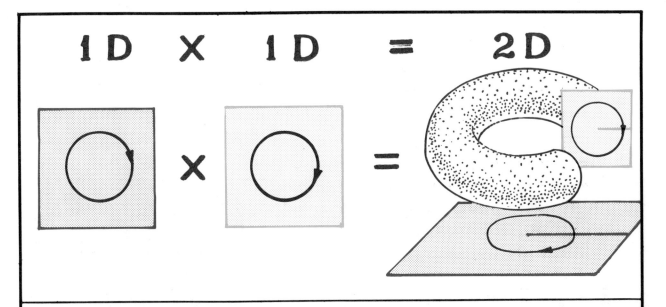

5.2.3. The geometric model for the states of this combined system is formed as follows: reduce the model for each clock (oscillator) to one dimension (a cycle), then take the Cartesian product of the two one-dimensional state spaces. The result is a two-dimensional torus, as explained in the preceding section.

In this *uncoupled* situation, there is *no entrainment.* Here's why.

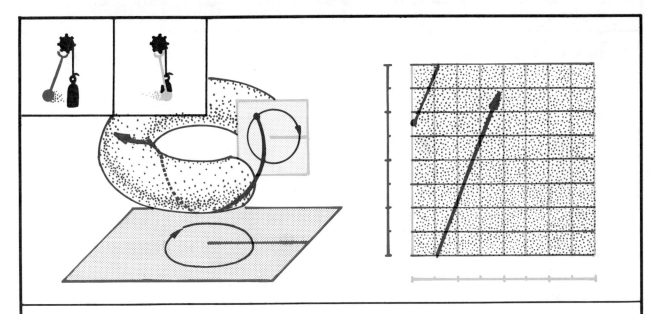

5.2.4. The trajectory of a point on the torus, corresponding to the time (phase) of each clock, winds around the torus. Assume the rate of each clock is constant. Then on the flat rectangular model of the torus, the trajectory is a straight line.

The slope of this line is the ratio: rate of second (green) clock to rate of first (blue) clock. This ratio, a constant, may be rational or not.

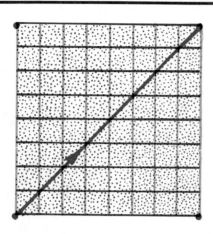

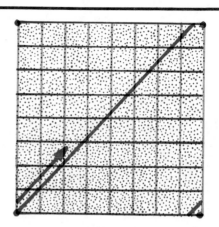

5.2.5. Ideally, if the two clocks run at the same rates, the ratio is one. Further, if they tell the same time, their phases are identical. Therefore, the trajectory on the flat torus is the diagonal line running from corner to corner.

5.2.6. If the system is slightly changed, by a speck of dust in the works of one of the clocks for example, the ratio of the rates changes. The slope of the straight line on the flat model of the torus changes slightly. It is no longer exactly equal to one. And the trajectory on the torus changes from a periodic trajectory to a solenoid, perhaps, or to a periodic trajectory which winds many times around, instead of just once.

This is the situation called *non-entrainment.* This means that *a slight change in the system results in a slight change in the ratio of the rates (frequencies) of the oscillators.*

Now we may explain *entrainment.*

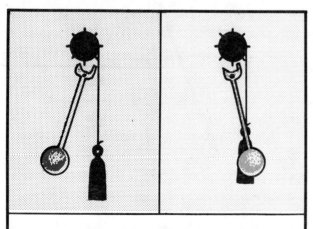

5.2.7. So far, we have assumed each clock is totally indifferent to the state of the other: the clocks are *uncoupled.*

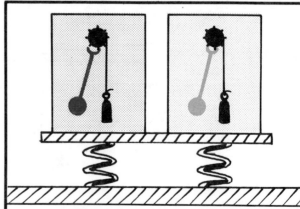

5.2.8. A slight mechanical connection between the two oscillators will create an interaction: the motion of each will influence the motion of the other.

Now the two oscillators are *coupled*. This means that the phase portrait is *perturbed* by the addition of small vectors at each point of the state space. This small vectorfield is added to the dynamical model representing the uncoupled system. Without saying exactly what this small perturbation is, one can conclude something about the coupled system anyway. This amazing conclusion is a geometric theorem of Peixoto, described at length in Volume 2 of this series.

Here is the geometric model for the same physical system, with coupling introduced between the mechanical oscillators.

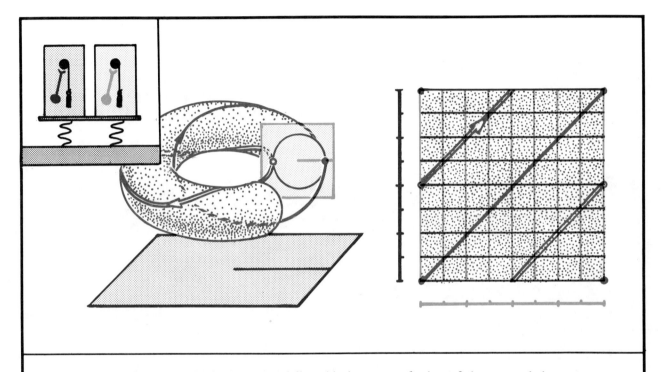

5.2.9. The dynamical system modeling this is a *perturbation* of the uncoupled system described previously. The theory of Peixoto ensures that for a typical perturbation, the perturbed model looks like this. There is a finite, even, number of closed trajectories. They all wind around the same number of times. Every other one is an attractor. The intermediate ones are repellors. There are no other limit sets. This kind of portrait is called a *braid*.

Even more, Peixoto's theorem says that this braid is *structurally stable*. This means that a small perturbation (another speck of dust) will make *no significant change in the phase portrait*. For example, a braid with periodic trajectories winding once around (equal rates of the clocks) will still be a once-winding braid after a small perturbation. Thus, Peixoto's theory provides a mathematical rationale for the *frequency entrainment* phenomenon observed by Huyghens. *Warning: The phases need not be entrained, only the frequencies.*

How do braids arise? Although Peixoto's theory provides a mathematical basis for this phenomenon, the actual mechanics of it is not yet clear. We will demystify the mechanics of braids in Section 5.4. First, we need to enlarge the geometric model from two dimensions to three. So, on to the next section.

5.3. THE RING MODEL FOR FORCED OSCILLATORS

We are now ready to apply the abstract ideas of coupled oscillators to some concrete examples. In this section (and the next two), we consider the coupling of mechanical oscillators. This is analogous to the work of Duffing described in the preceding chapter. But there, the forcing oscillator (motor) was coupled to a system which wished to come to rest (damped pendulum). Here, we couple the forcing oscillator (motor) to a system which tends to a self-sustained oscillation (clockworks). This clockworks will be assumed to be *affected by the coupling*, while the turntable motor is *unaffected.*

5.3.1.   The turntable motor is so well regulated that its speed, once set with the control knob, is unaffected by the load. The forcing oscillation is coupled to the clock pendulum by a light spring. The stiffer the spring, the greater the effect of the driving oscillation on the periodic motion of the clock pendulum.

The most useful geometric model for the state space of this system is three-dimensional. It is essentially the same as the *ring model* of section 4.1. We will place the torus in this three-dimensional context.

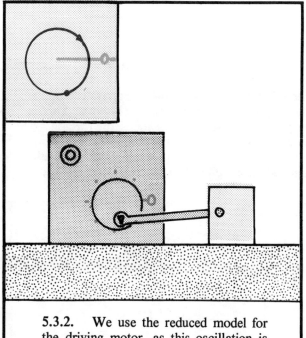

5.3.2. We use the reduced model for the driving motor, as this oscillation is unchanged by the coupling. Only its *phase* is important.

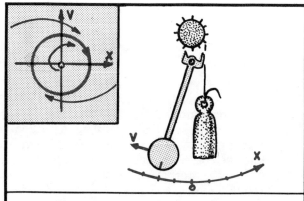

5.3.3. But the state space of the clockworks will be visualized as a two-dimensional phase plane. The phase portrait consists of a periodic attractor, representing the self-sustained oscillation of the pendulum. The width of this cycle then corresponds to the *amplitude, A,* of the oscillation of the pendulum of the clock. This amplitude will be affected by the force of the driving motor, communicated through the coupling spring.

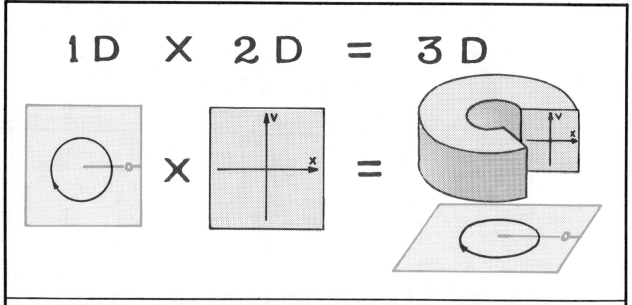

5.3.4. The Cartesian product of the clock plane and the motor cycle is the three-dimensional *ring model* for the combined system. (Compare with 5.2.3.)

This scheme represents a special case of coupled systems, in which there is a dominant partner and a more flexible one. We have used the same strategy in the ring model for the forced pendulum (see Section 4.2). A more general scheme would use two dimensions for each oscillator, for a total of four dimensions.

The regulated motor maintains its rate, but the motion of the pendulum is affected by the periodic force of the spring. To see how it is affected, we may take out the coupling spring, then replace it.

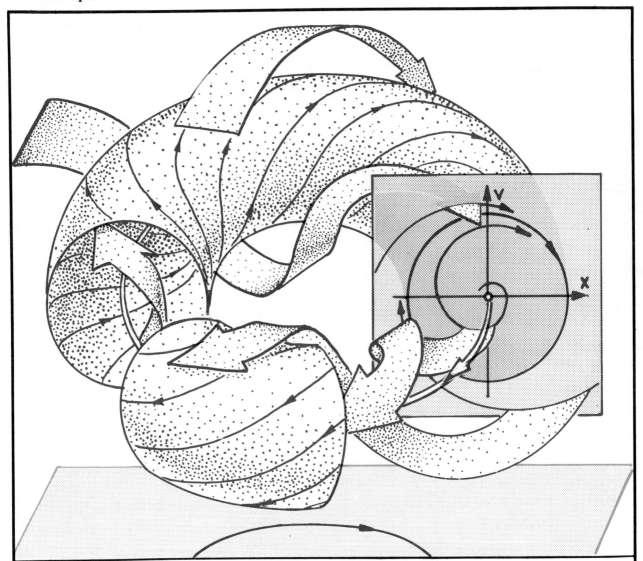

5.3.5. With no spring, the two oscillators are uncoupled. Once the transients die away, the motion may be recorded as a trajectory on a torus, as explained in Section 5.1. Unlike the green torus in the ring model of the forced pendulum in Chapter 4, this red torus is an *invariant manifold* of the dynamical system. This means that every trajectory which begins on the torus, stays on the torus. In fact, the red torus is *attractive*, yet probably *not an attractor*. It is *attractive* in this sense: if we put the combined system in an initial state *off the torus* (for example, by giving the pendulum a shove to a larger amplitude) and let it go, the resulting trajectory will be attracted to the torus, as the amplitude decays to the original value.

And it may be *not an attractor* in the following sense: if we put the combined system in a typical initial state *on the torus* and let it go, the resulting trajectory may be a periodic trajectory on the torus. But when we say a set is an attractor, we mean not only that it is attractive, but also that it is *transitive:* that is, most trajectories on it wander all over it.

Thus, not all attractive tori are attractors.

When we reattach the spring, the two oscillators are coupled. The phase portrait is a perturbation of the picture described above for the uncoupled system. According to an important theorem of mathematical dynamics, the perturbed portrait still has an attractive, invariant torus. If the perturbation (coupling) were gradually turned on, the uncoupled torus would be gradually deformed into the coupled one. Therefore the theory of Peixoto applies, as described in the preceding section. The phase portrait contains a braid of periodic attractors on the torus. The invariant torus is attractive, yet not an attractor. The braided periodic trajectories within the torus are the actual attractors, in the coupled case.

Because of these braids, the clockworks and the motor are *entrained in frequency, but not in phase.* In the next section, we will examine the dynamics of these braids.

5.4. BRAIDS: THE DYNAMICS OF ENTRAINMENT

In Section 5.2, we encountered the braided periodic attractors in the two-dimensional torus model for coupled oscillators. To explain the mechanics of frequency entrainment in our forced mechanical oscillator, we will reconsider the braids in the three-dimensional ring model of the preceding section.

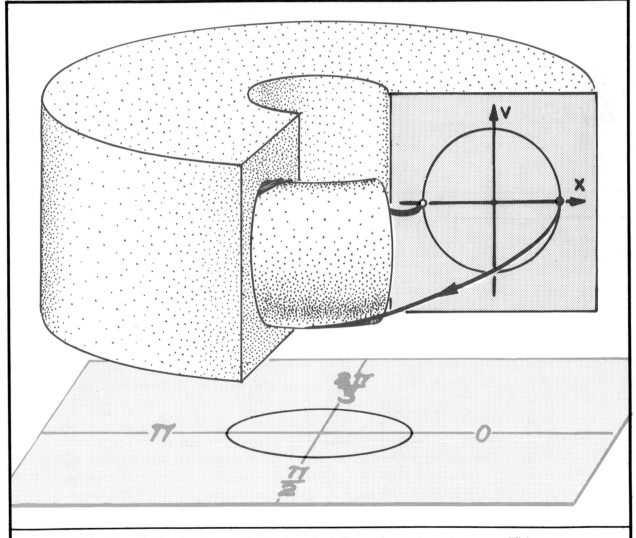

5.4.1. Recall that the ring model for forced oscillators has an invariant torus. This attractive torus corresponds to the torus (reduced) model of Section 5.2. Now we put the 2D torus back into the context of the 3D ring model. By performing armchair experiments with our mechanical device, we will see how braids arise on this torus.

In this series of experiments, we will couple the turntable motor to the pendulum of the clock-works with a very feeble spring. Thus, the phase portrait of the combined system will be a slight variation of the uncoupled system, in the ring model. The uncoupled system was shown in 5.3.5. Also, we will begin by setting the speed of the driving motor to the natural frequency of the clock. This is the situation called weakly-coupled, *isochronous* oscillators.

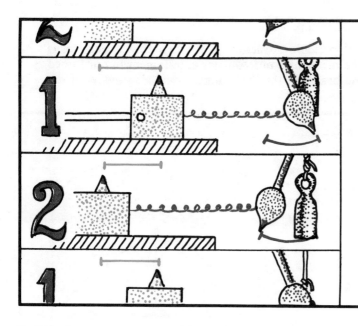

5.4.2. Start up the motor. After a brief start-up transient, we observe the driven weight and the pendulum swinging *in phase*. The motion of the pendulum has been influenced by the motion of the weight, through the intercession of the weak coupling spring. We suppose the weight has a bigger swing than the pendulum, as shown here. Then the only result of this influence, in contrast to the uncoupled situation, is a slight increase in the amplitude (width of swing) of the pendulum.

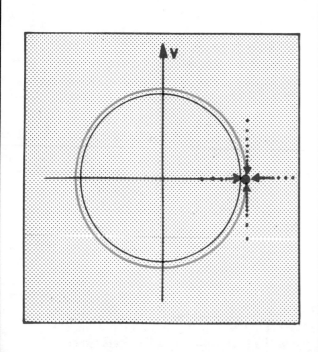

5.4.3. Now let's look at the trajectory of this motion in the ring model. First, we slice the ring in the strobe plane corresponding to phase zero of the driving oscillator. In this slice, the invariant torus of the uncoupled system shown in 5.4.1 appears as a red circle. Because the amplitude of the pendulum has been slightly increased by the coupling, the green locating torus (not invariant) for this trajectory is slightly fatter than the uncoupled red torus. Thus, in this strobe plane, it appears as a slightly larger circle. The trajectory of the isochronous, in-phase, periodic motion observed in the preceding panel winds around a green locating torus, and appears in the strobe plane as a point, shown here as a *solid red dot*. The blue dots represent trajectories which are approaching the attractor.

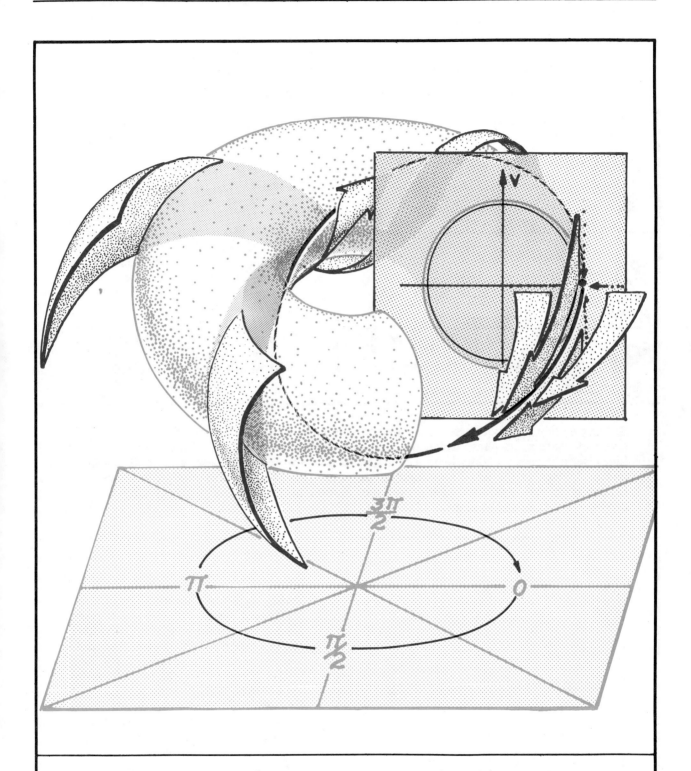

5.4.4. Here is the isochronous in-phase trajectory in the ring model. Note that the attractive trajectory goes through phase zero of the driving motor at roughly the same time it passes phase zero of the driven clock. It is *in-phase*. Also, it is *isochronous*, as the frequencies are entrained to be equal. This periodic trajectory is actually an *attractor*.

Recall that in Section 5.2, Peixoto's theory predicted that in the torus model, we would find not only a periodic attractor, but a periodic repellor as well. Another experiment, with the motor and clock isochronous but out-of-phase, will locate this periodic motion in the ring model.

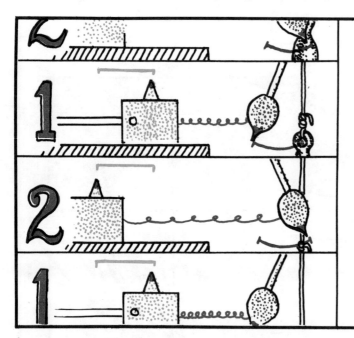

5.4.5. In this experiment, the clock pendulum is set at phase π (half-cycle, or full left) and the driving weight at phase zero (full right). They are out-of-phase when the experiment begins. They are as close together as they can get. We let them go, and the start-up transient dies away. After half a cycle, in step 2, they are as far apart as they can get. Here the clock's pendulum is opposed by the full force of the stretched coupling spring. After a full cycle, step 1 repeats, and they are closest together at the same time again. Once more, the clock's pendulum is opposed by the spring's greatest force. The effect of the coupling is to reduce the amplitude of the pendulum's motion.

But this motion does not persist. Its trajectory is not an attractor. for if the clock began slightly ahead of phase π, its phase would drift forward until it were in-phase with the driving weight. Likewise, if it began slightly behind phase π, its phase would drift backwards until it were in-phase. In fact, in the two-dimensional torus model (panel 5.2.9), this motion was a repellor, belonging to the separatrix. Let's plot this motion in the ring model, beginning with the driving-phase-zero strobe plane.

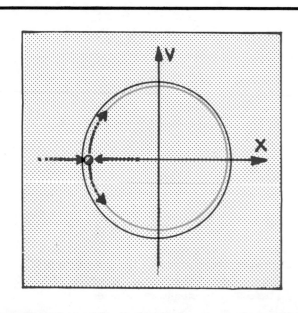

5.4.6. Again, the invariant torus of the uncoupled system cuts the strobe plane in a circle, shown here in red. Another smaller torus, not invariant but convenient for visualization, cuts the plane in the green circle. The periodic trajectory corresponding to the isochronous, out-of-phase oscillation winds once around the green, locating torus. It cuts the strobe plane in a single point, shown here as a *half-filled red dot*. Note that this trajectory *repels* in the direction along the torus, and *attracts* in the perpendicular direction. The repellor in the torus model becomes a saddle in the ring model.

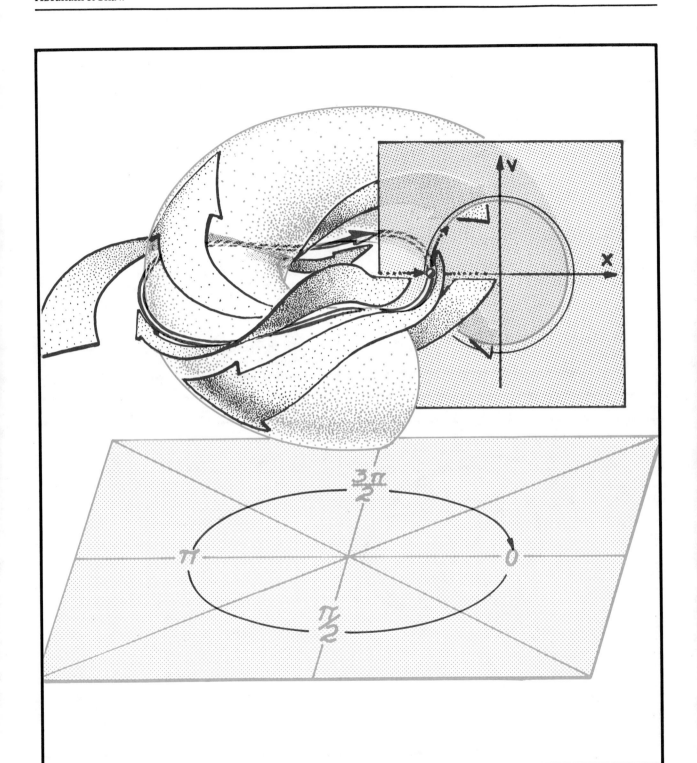

5.4.7. And here is the strobe plane, placed in the three-dimensional ring model. The red trajectory records the motion of the ideal experiment, with a perfectly out-of-phase, iso-chronous oscillation. This periodic trajectory is not actually an attractor, but a *periodic saddle*. It attracts amplitudes but repels phases, as shown by the ribbon arrows.

Now we have located the braided cycles, and we are ready to determine the shape of the invariant torus they are braided around.

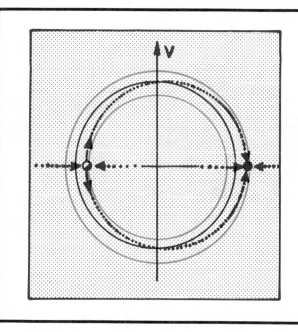

5.4.8. The invariant torus of the weakly coupled system is close to that of the uncoupled system. It is composed of the outset of the periodic saddle. Here, we show this outset in red, as cut by the strobe plane. Note that it lies between the green locating tori used in the preceding constructions.

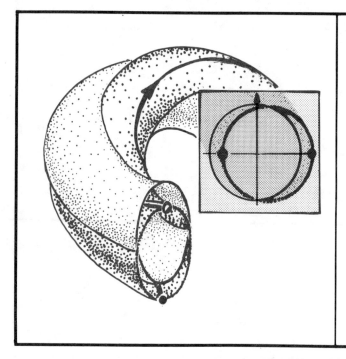

5.4.9. Here, in three dimensions, is the new invariant torus of the weakly coupled system (coarse shading). It is close to the old invariant torus of the uncoupled system (fine shading). The periodic attractor and the periodic saddle wind around the new torus. Both are isochronous. That is, they wind once around the waist of their invariant torus.

Another important feature of the phase portrait of the combined system is the *central repellor*. There is a repellent periodic trajectory near the center of the ring model. It corresponds to a small oscillation of the clock pendulum, out of phase with the driving motor. It is an *unstable equilibrium*. That is, it is possible in principle to balance the clock precariously in this mode of oscillation, just as it is possible to balance a pendulum at the top of its swing (see Section 2.1).

5.4.10. Again, we imagine a green, noninvariant, locating torus. This one is quite thin, and lies near the center of the invariant torus. Here, we show the strobe plane. It cuts the locating torus in the small green circle, and the invariant torus in the dotted red circle. The isochronous central repellor winds once around the green locating torus, meeting the strobe plane in a single point, shown here as a *hollow red dot*.

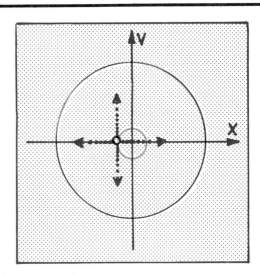

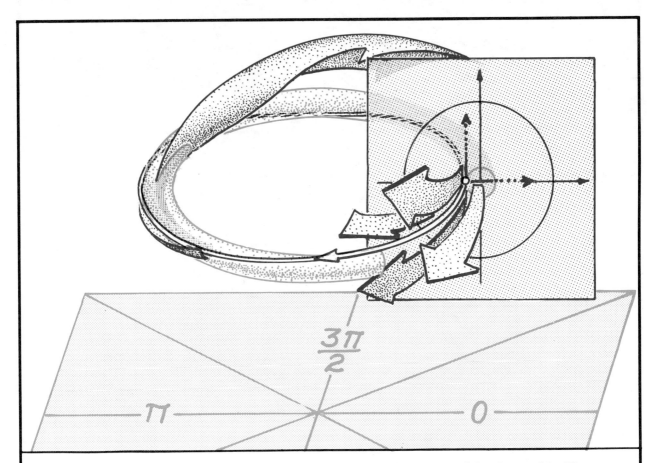

5.4.11. Here is the strobe plane, in the three-dimensional context of the ring model. The outset of the periodic saddle comprises the invariant torus. The outset of the central repellor comprises the central portion of the basin of the periodic attractor.

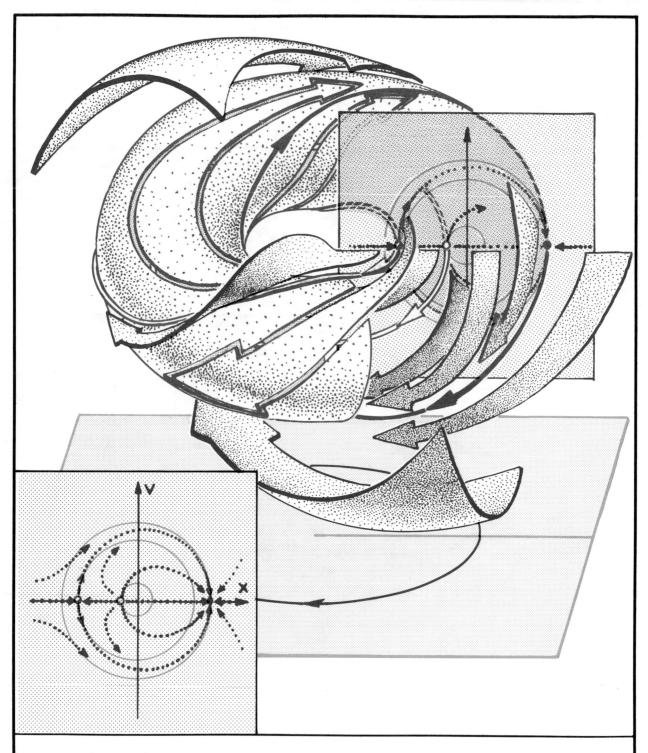

5.4.12. Putting all the pieces together, we see the two braided periodic trajectories on the invariant (red) torus, with the central repellor on its locating (green) torus within. The inset shows the driving-phase-zero strobe plane.

Now we see the braid: one periodic attractor and one periodic saddle, each isochronous, occupy the perturbed invariant torus. The attractor represents an in-phase oscillation of the coupled system. This situation is sometimes called *phase entrainment*. But unlike frequency entrainment, it is not *structurally stable*. This means that a further perturbation of the system (for example, a slight change in the speed of the driving motor) may shift the phase difference of the periodic attractor away from zero, while the clock frequency will still be entrained by the motor. Worse, for two general oscillators without an obvious zero phase, phase entrainment does not even make sense.

We have established that the phase portrait of the compound oscillator has at least three periodic trajectories: the braided saddle and attractor, and the central repellor. These three motions are isochronous. But we have only experimented with approximately equal frequencies. What happens if we really change the speed of the driving motor?

5.5.   RESPONSE CURVES FOR FREQUENCY CHANGES

To see how this phase portrait depends on the speed of the driving motor, we must repeat the experiments of the preceding section many times, with different driving frequencies.

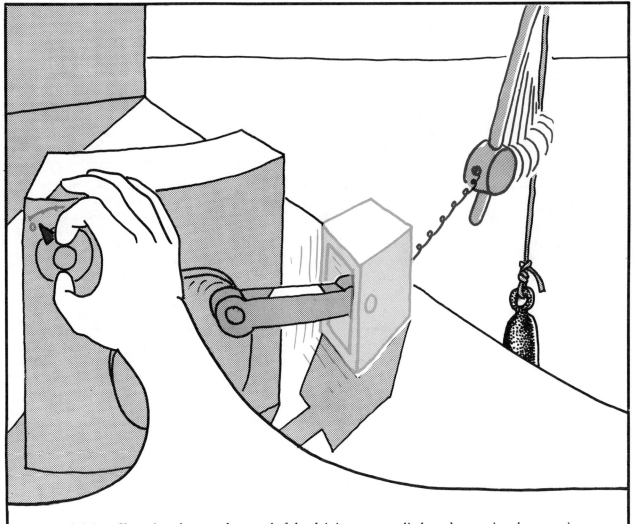

5.5.1.   Changing the speed control of the driving motor a little and repeating the experiment, we will observe almost identical results: three isochronous periodic motions of the compound system. The pendulum is still isochronous, or entrained, with the driving frequency. Yet there are subtle differences in the *amplitudes* of the three isochronous periodic motions. So, we must carefully measure these amplitudes in our experiments.

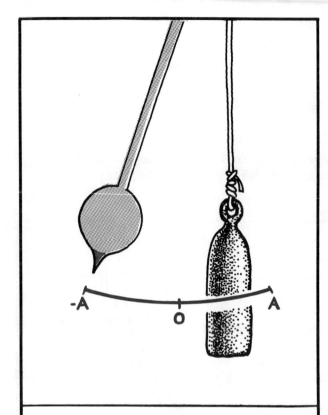

5.5.2. The amplitude is measured along the arc of the swinging pendulum. It is the maximum angle attained by the bob, in deviation from the vertical. It is a positive number.

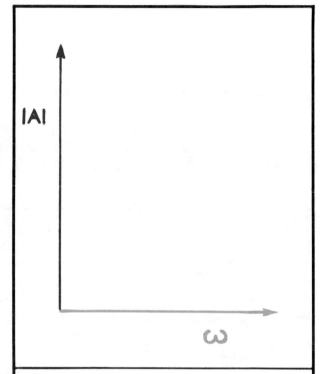

5.5.3. In each experiment with the forced oscillator, we will measure only the amplitude of the periodic motion found, called the *response amplitude,* and the frequency of the driving motor, or *driving frequency.* These two measurements may then be plotted in this *response plane,* of response amplitude versus driving frequency.

After many experiments, we will obtain the graphs of these response amplitudes, regarded as functions of the driving frequency. These *response curves* comprise the *response diagram,* also called the *bifurcation diagram of the one-parameter system.*

We have already done the isochronous experiments, and may find the response amplitudes by looking back at the strobe plane illustrations of the preceding section. Recall that we use the following *strobe plane convention:*

    solid dot   = attractor
    half-filled dot = saddle
    hollow dot   = repellor

for representing the three trajectories as points in the strobe plane.

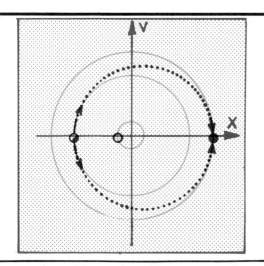

5.5.4.   Here is the strobe plane portrait for the isochronous case. The diameter of the locating torus, represented as a green circle here for each of the three periodic trajectories, is the response amplitude.

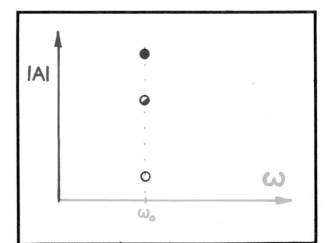

5.5.5.   Here, the three response amplitudes are plotted above the point on the horizontal axis, $\omega 0$, representing the driving frequency. This corresponds to the speed of the motor, which in this case is isochronous with the clock (60rpm).

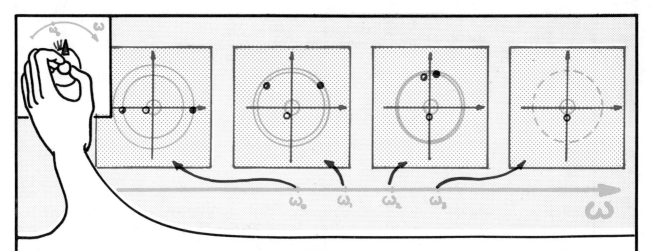

5.5.6.   Increasing the frequency of the driving oscillation, ω, the isochronous oscillation of the clock pendulum (periodic attractor, solid dot in the preceding plot) lags behind the driving oscillation. Simultaneously, the phase of the unstable mode (periodic saddle, half-filled dot in the preceding plot) advances ahead of the driver. As the the driving frequency increases further, the amplitudes of these two periodic trajectories become closer to each other. Eventually, they coincide, cancel, and disappear. This is a catastrophic change in the phase portrait, which now has only one isochronous periodic trajectory, instead of three.   And the remaining one is repelling!   In this illustration, the amplitudes and phases of the periodic trajectories are shown in the cross-section of the ring model corresponding to phase zero of the driving oscillation. The sequence of sections, from left to right, corresponds to increasing driving frequencies, $\omega 0$, $\omega 1$, $\omega 2$, and so on.

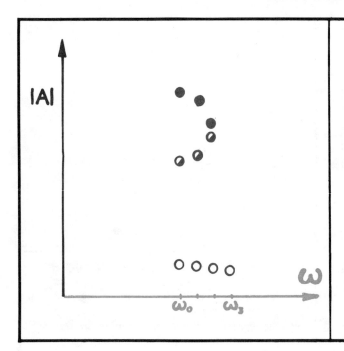

5.5.7. Measuring the response amplitudes of the periodic motions, seen as the diameters of the green circles in each of the strobe planes in the preceding panel, we record the results of the series of armchair experiments in the response plane. Over each of the chosen driving frequencies, we record the observed response amplitudes.

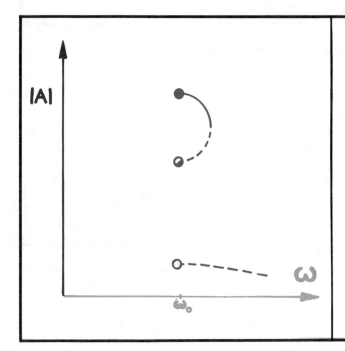

5.5.8. After more experiments if necessary, we can fill in the complete response curves for the isochronous motions. The change from three motions to only one turns out to be a smooth one, as two curves join in a parabolic shape.

Here we have used the *response plane convention:* solid curves are the tracks of attractors, dotted curves are the tracks of saddles or repellors.

Known as the *dynamic annihilation catastrophe,* this particular bifurcation diagram is an example of *bifurcation behavior,* the subject of *Volume 3* of this series.

A similar event results from *decreasing* the speed of the driving motor, as we shall see in the next illustration.

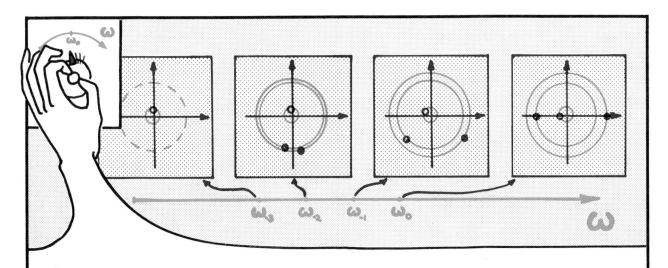

5.5.9. Decreasing the driving frequency results in a similar bifurcation event, as shown in this sequence of sections. The phase of the attractor advances, while the phase of the saddle lags behind. Eventually, they meet and wipe each other out. The sequence of sections shown here, *from right to left,* corresponds to decreasing driving frequencies, $\omega 0$, $\omega 1$, $\omega_{-}2$, and so on.

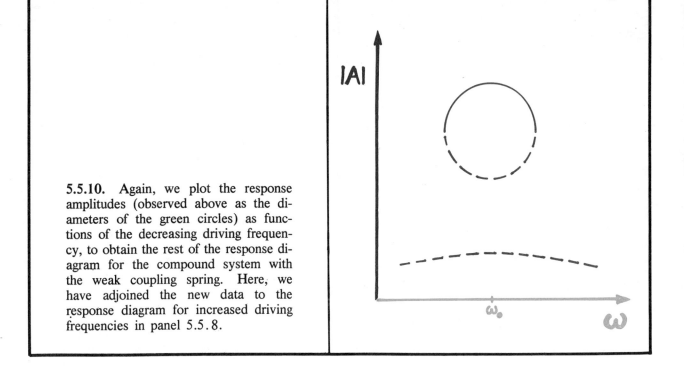

5.5.10. Again, we plot the response amplitudes (observed above as the diameters of the green circles) as functions of the decreasing driving frequency, to obtain the rest of the response diagram for the compound system with the weak coupling spring. Here, we have adjoined the new data to the response diagram for increased driving frequencies in panel 5.5.8.

On the other hand, we could change the spring and repeat all of the experiments. For example....

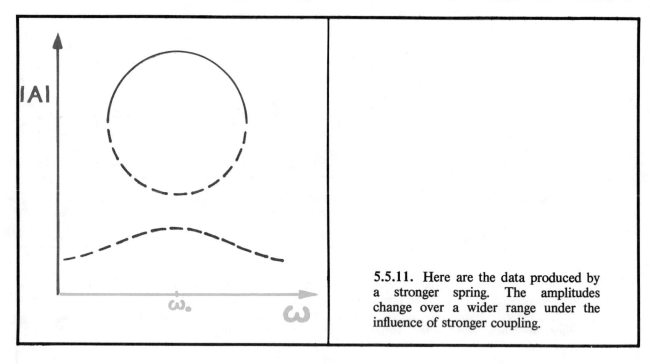

5.5.11. Here are the data produced by a stronger spring. The amplitudes change over a wider range under the influence of stronger coupling.

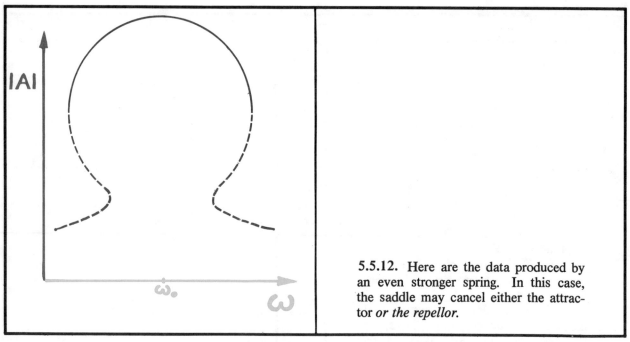

5.5.12. Here are the data produced by an even stronger spring. In this case, the saddle may cancel either the attractor *or the repellor.*

These response data record amplitudes of oscillation, as functions of *two control parameters:* driving frequency and coupling strength. Stacking up a number of response diagrams such as the two preceding ones, for stronger and stronger springs, would produce a three-dimensional plot. This is known as the *bifurcation diagram of the two-parameter system.* Many examples will be described in Volume 3 of this series. A more compact representation of the same data may be made, just by superimposing the plots on the same planar diagram. This is called the *response diagram* of the two-parameter system.

Here is the response diagram for the two-parameter system. The parameters are coupling strength and driving frequency.

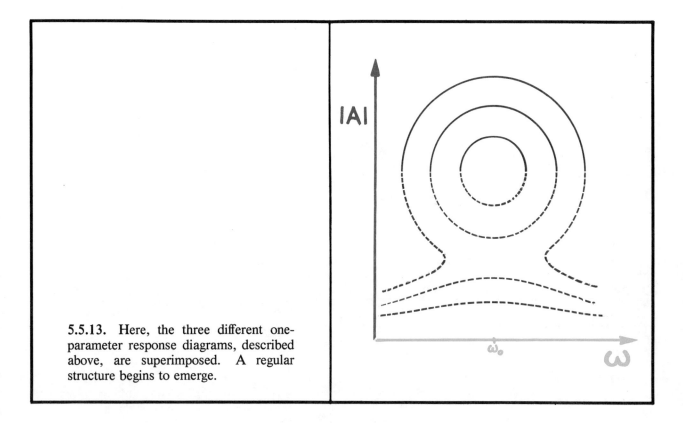

5.5.13. Here, the three different one-parameter response diagrams, described above, are superimposed. A regular structure begins to emerge.

This diagram tells almost everything about the amplitudes of the isochronous oscillations of an ideal, self-sustained oscillator, when driven by another, regulated oscillator. Many more oscillations, called harmonics, are known to occur in such systems, however. Their discovery and analysis was greatly facilitated by the development of electrical oscillators, to which we now turn.

5.6. FORCED ELECTRICAL OSCILLATORS

Lord Rayleigh's model for self-sustained oscillations in organ pipes and violin strings (Chapter 3) applies very naturally to vibrations in an electrical context. In fact, Rayleigh himself carried out the application to an electrical vibrator invented by Helmholtz. Forty years later, the early electrical engineers found that Rayleigh's model worked very well for the vacuum tube oscillators used in the first radio transmitters. Further, Rayleigh had briefly studied a coupled system of two such oscillators —a large one forcing a small one. But real progress in understanding coupled oscillators awaited the further development of radio frequency electronics. In 1921 the time was right, and Van der Pol began this progress. In this section, we describe his program, which culminates this Volume.

Electronic oscillators are easier for experimentalists to manipulate than clocks, but harder for onlookers to understand.

5.6.1. First, we replace the clockworks with an electric oscillator. Not an electric clock, but a radio frequency oscillator running millions of times faster. (See Section 3.4.)

5.6.2. As the current and voltage fluctuations are much too rapid to be observed on the panel meters, we connect an oscilloscope to the device. Horizontal deflection measures current, and vertical measures voltage, of the output circuit of the oscillator.

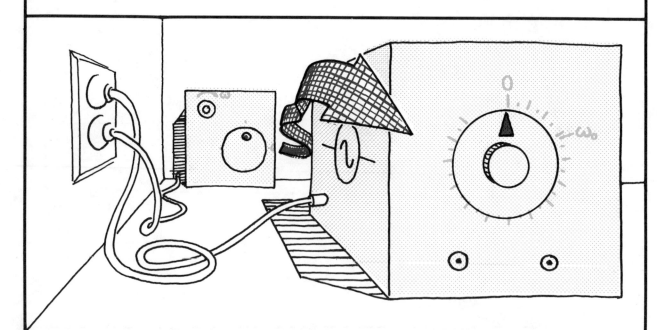

5.6.3. Next, we replace the regulated driving motor with another electrical oscillator. This is a VFO (variable frequency oscillator), a well regulated sinusoidal oscillator with a frequency adjustment knob.

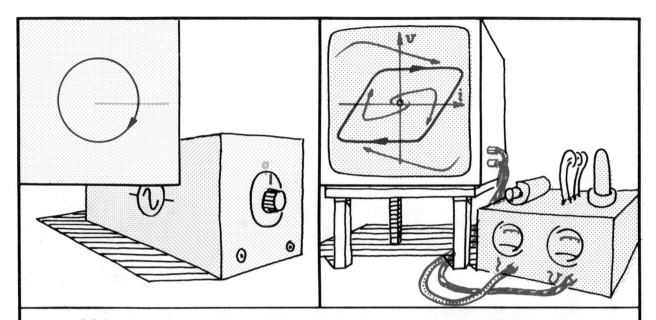

5.6.4. As in the preceding section, the state space for the driving oscillator is a circle. One dimension, phase of the driving oscillation, suffices. The state space of the driven oscillator is a plane. Before coupling, the phase portrait consists of a single periodic attractor, cycling around a point repellor.

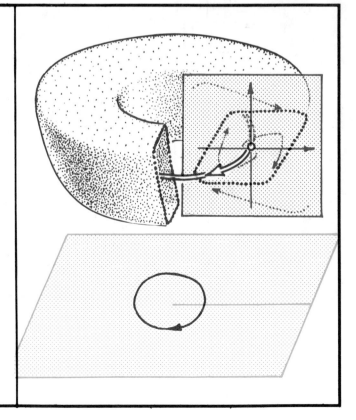

5.6.5. The combined state space is again a *ring model,* the Cartesian product of the circle and the plane. Before coupling the two oscillators, the combined phase portrait consists of an attractive, invariant torus around a periodic repellor.

5.6.6. Next, the coupling. The output of the driving generator is connected to the input of the driven oscillator. The actual coupling is electromagnetic, through an auxiliary coil around the inductor of the driven oscillator.

5.6.7. Finally, we will utilize a pulse from the driving oscillator. This brief pulse is emitted each time the driving oscillation passes phase zero. It is connected to the high voltage power supply of the oscilloscope, so that the beam is on only during the brief pulse. This scheme replaces the strobe lamp of the mechanical device in the preceding section. Thus, we see on the screen of the oscilloscope only the cross-section of the ring model corresponding to driving phase zero.

Let's try out this setup, which is essentially the same as the apparatus in Van der Pol's laboratory. First, we unplug the coupling wire between the two oscillators. Then...

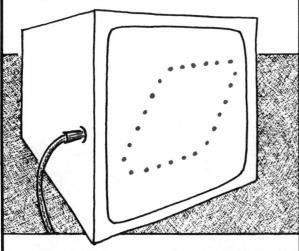

5.6.8. Turning on everything, we see a dotted outline of the periodic attractor of the driven oscillator on the oscilloscope screen. These dots are trajectories crossing driving-phase-zero strobe plane. The oscilloscope is strobed by the zero-phase pulses from the driving oscillator.

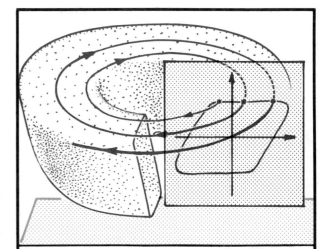

5.6.9. Prolonging this cross-section around the ring model, we see that the trajectory is a periodic attractor or a solenoid. The way it winds around the attractive (red) torus depends on the relationship between the two frequencies, as described in section 5.1.

Before plugging the coupling wire back in, we can try changing the frequency of the driving oscillator.

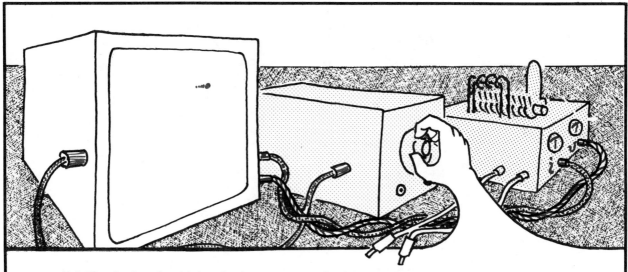

5.6.10. Setting the driving frequency to exactly the same frequency as the driven oscillator would reduce the trajectory to a single dot, in the cross-section of the attractive torus. But it is impossible to set it exactly right, so the dot wanders around.

Now we plug in the coupling wire between the two oscillators.

5.6.11. Miraculously, the wandering dot settles down in a single spot. Further tiny movements of the driving frequency knob may move this spot back and forth on the driven (red) cycle, but at any one frequency it does not wander.

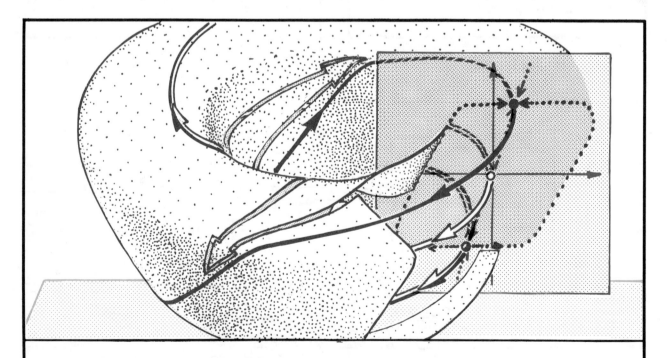

5.6.12. But we expect this, after the explanations of the preceding section. The *structural stability* of the coupled phase portrait implies *frequency entrainment* of the coupled oscillators.

For example, here is one called the third sub-harmonic (see section 5.2). It may be observed with Van der Pol's original apparatus, by setting the driving frequency to three times the frequency of the driven oscillator.

5.6.13. Set the frequencies, then turn on the oscillators, and connect the coupling wire. You will see three dots on the red cycle.

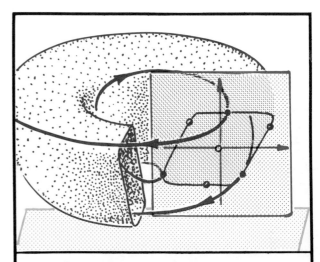

5.6.14. These three dots, prolonged around the attractive (red) torus, trace a periodic attractor winding thrice around.

A few zillion experiments like these will reveal further details of the response diagram, as we mentioned at the end of the preceding section.

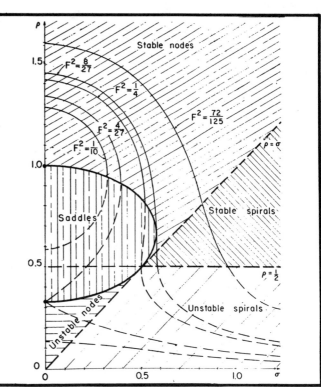

5.6.15. Here is the response diagram, as described by Stoker [14] in 1951. Unexpected exotic details would be discovered a decade later.

And for these exotic details, of chaotic attractors and tangled separatrices, turn to Volume 2 of this series!

CONCLUSION   The hardy reader, arriving at this point without extensive short cuts, has seen the dynamics of the classical period, 1600-1950. In fact, this subject had a full and happy life, and died of natural causes. The masterly text of Stoker (14) is the eloquent epitaph of *classical dynamics*. Fortunately for the evolution of our species on planet Earth, this demise was foreseen in 1882 by Poincare'. His vision, nourished in secrecy by his faithful followers in Russia and America, emerged in the 1950's into the mainstream of mathematics, united with the outstanding applications of the classical period, and acquired new electronic laboratory tools. Thus, *modern dynamics* was born.

Our goal in this book, the first of a series, has been not only to survey classical dynamics, but to do so from the viewpoint of the modern period. Thus, the groundworks for the following volumes have been laid. The next volume will report on the amazing discoveries of the modern period: *generic tangles* and *attractive chaotics*.

Basically, the classical period ended because the analytical techniques were exhausted. The modern period relies upon the geometric techniques pioneered by Poincare'. But simultaneously, experimental dynamics was revolutionized by electronic inventions. With fast analog and digital computers replacing the mechanical apparatus of the classical dynamics laboratory, amazing discoveries were rapidly made. The theory and experiments evolved together in the recent decades, and a new set of paradigms emerged to revolutionize the sciences. This story is the subject of the next two volumes of our series.

APPENDIX: SYMBOLIC EXPRESSIONS

As we have freely admitted in the preface, our presentation of mathematical ideas by visual and verbal representations alone, without rigorous, symbolic expressions, is incomplete. It is unfair to mathematics and to the mathematically literate reader. But, wanting to give the maximum understanding to readers without experience of symbolic expressions, we decided to collect all the symbols in a closet of their own. This is it.

Our goal is provide symbolic expressions for the basic concepts of dynamics (section A), and for the main examples of dynamical systems *(section B and Disk One).*

A. BASIC CONCEPTS

In this section, we complete the representation of the basic concepts from Chapter 1 by adjoining their analytical (formal, symbolic) definitions. This is only a summary. For more details, consult the texts listed in the Bibliography.

State spaces are smooth manifolds. The examples used in the book are:

$R = R^1$, the real number line,
$R^n = R \, x...x \, R \, (n \ times)$, Euclidean n-space,
$S^1 = T^1$, the *circle*,
$T^n = S^1 \, x...x \, S^1 \, (n \ times)$, the *n-torus*,

and their Cartesian products, the various cylinder spaces.

In such a manifold, M, the space of tangent vectors at a point, $x \varepsilon M$, is naturally identified with Euclidean space, $T_x M = R^n$. Thus, a smooth vectorfield on M, is represented by a smooth mapping,

$$V: M \to R^n; \; x \to (V_1(x),..,V_n(x))$$

and a smooth curve, *c: $R \to M$ is an integral curve* or *trajectory* of the vectorfield, V, if for all t, $c'(t)=(c(t), V(c(t)))$, where $c'(t)$ is the tangent vector to c at t. In case $M=R^n$, the tangent vector is $c'(t)=(c(t), v(t))$, where $v(t)$ is the velocity vector, $v(t)=lim \; [c(t+h)-c(t)]/h$ as h goes to zero. Thus, the curve, c, is an integral curve if

$$c(t) = (c_1(t),...,c_n(t)),$$

and the component functions satisfy the system of ordinary differential equations of first order,

$$c'_1(t) = V_1(c_1(t),..,c_n(t))$$

$$\vdots$$

$$c'_n(t) = V_n(c_1(t),..;c_n(t))$$

Here c'_i denotes the ordinary derivative.

Two special cases were described in Section 1.3. First, if $V(x)=0$, then x is a *critical point* of the vectorfield, and the constant curve,

$c: R \rightarrow M; t \rightarrow c(t)=x,$

is an integral curve.

Second, if c is a non-constant integral curve and $c(t+T)=c(t)$ for some (smallest positive) real number T, then c is a *periodic trajectory of period T*.

An integral curve is *complete* if it is defined for all real numbers. Suppose c is a complete integral curve at $x=c(0)$. The point x is called the *initial point* of c. The *omega-limit set* of x is the countable intersection,

$$\omega(x)=\cap\{cl(c[n,\infty)) \mid n\varepsilon N\}$$

where N denotes the positive integers, $[n,\infty)$ denotes the closed ray, and $cl(A)$ denotes the closure of the set A. Similarly, the *alpha-limit set* of x is the countable intersection,

$$\alpha(x)=\cap\{cl(c(-\infty,-n]) \mid n\varepsilon N\}$$

Let $L \subset M$ be a subset. Then the *inset of L* is the set

$$In(L)=\{x\varepsilon M \mid \omega(x) \subset L\}$$

and similarly, the *outset of L* is the set

$$Out(L)=\{x\varepsilon M \mid \alpha(x) \subset L\}$$

If L has an open neighborhood U, in its inset, $L \subset U \subset In(L) \subset M$, then L is *attractive* (there are many important variants of this definition) and In(L) is its *basin of attraction*. An *attractor* is more than an attractive set. It is a subset, $A \subset M$, which is attractive, and which has *no proper subset* which is attractive. The *separatrix* is the complement of all basins of attraction:

$$Sep=\{x\varepsilon M \mid \omega(x) \text{ is not contained in an attractor}\}$$

B. EXEMPLARY SYSTEMS

Using the notational conventions described in the preceding section, we give here the formulary of the examples described in the text, in tabular form.

Example: 1.
Section: 1.6.
Type   : Negradient system.
Origin : Newton, ca. 1665.
Space : Plane, R^2
System :

$$x' = x^3 + x^2 - 2x$$
$$y'=-y$$

Remarks: Potential function,

$$F(x) = x^4/4 + 2x^3/3 - 3x^2/2 + y^2/2.$$

Example: 2a.
Section: 2.1.
Type   : Simple pendulum.
Origin : Newton, ca. 1665.
Source : Stoker [14], p. 61.
Space : Cylinder, $S^1 x R$.
Coords : angle of elevation, A.
        angular velocity, B.
System :

$$A' = B$$
$$B' = F sin(A) - cB$$

Remarks: coefficient of viscous damping, c,
        weight of pendulum, F.

Example: 2b.
Section: 2.2.
Type   : Buckling column.
Origin : Stoker, 1951.
Source : Stoker [14], p. 54.
Space : Plane, R^2.
Coords : displacement, x.
        velocity, y.
System :

$$x' = y$$
$$y' = (-1/m \ [a_3 x^3 + a_1 x + cy]$$

coefficients, $$a_1 = A + C - 2P/l$$
                 $$a_3 = B + D - P/l^3$$

Remarks: coefficient of viscous damping, c
      mass, m
        length or column, 2l
        vertical force, P
        restoring force of primary (lateral) hard spring, $Ax + Bx^3$
        restoring force of secondary (hinge coil) hard spring, $Cx + Dx^3$

Example: 2c.
Section: 2.3.
Type   : Spring.
Origin : Rayleigh, 1877.
Source : Stoker [14], p. 15.
Space : Plane, R².
Coords : displacement, x.
          velocity, y.
System :

$$x'=y$$
$$y'=(-1/m)\ \{a_3x^3+a_1x+cy\}$$

Remarks: mass, m
          restoring force of spring, $a_1x+a_3x^3$, $a_1>0$,
          hard spring, $a_3>0$
          linear spring, $a_3 = 0$
          soft spring, $a_3<0$

Example: 3.
Section: 2.4.
Type   : Predator-prey.
Origin : Volterra, Lotka, 1924.
Source : Hirsch and Smale, [5], p. 259.
Space : Plane, R²
Coords : prey population, x.
          predator population, y.
System :

$$x'=(A-By)x$$
$$y'=(Cx-D)y$$

Remarks: A,B,C,D>0
          saddle point, (0,0)
          center, (D/C, A/B)

Example: 4a.
Section: 3.1-3.
Type   : Self-sustained oscillation.
Origin : Rayleigh, 1883.
Source : Stoker, [14], p. 119.
Space : Plane, R²
Coords : current, x.
          voltage, y.
System :

$$x'=y$$
$$y' = (-1/CL)\ \{x+By^3-Ay\}$$

Remarks: capacitance, C>0
        inductance,   L>0
        characteristic function of vacuum tube, Bv^3-Av, A,B>0

Example: 2d.
Section: 4.3.
Type   : Forced spring.
Origin : Duffing, 1908.
Source : Stoker [14], p. 81.
Space : Ring, R^2xS^1.
Coords : displacement, x.
        velocity, y.
        driving phase, θ
System :

$$x'=y$$
$$y' = (-1/m) \{a_3x^3+a_1x+cy\}+Fcos(\theta)$$
$$\theta'=\omega$$

Remarks: mass, m
        restoring force of spring, $a_1x+a_3x^3$, $a_1>0$,
        hard spring, $a_3>0$
        linear spring, $a_3 = 0$
        soft spring, $a_3<0$
        coupling strength, F
        driving frequency, ω

Example: 4b.
Section: 5.4.
Type   : Forced, self-sustained oscillation.
Origin : Van der Pol, 1922.
Source : Stoker, [14], p. 147.
Space : Ring, R^2xS^1.
Coords : current, x.
        voltage, y.
        driving phase, θ
System :

$$x'=y$$
$$y' = (-1/CL) \{x+By^3-Ay\} +Fcos(\theta)$$
$$\theta'=\omega$$

Remarks: capacitance, C>0
        inductance, L>0
        characteristic function of vacuum tube, Bv^3-Av, A,B>0
        coupling strength, F
        driving frequency, ω.

NOTES

Preface

[1] Dynamics, a visual introduction, in F.E.Yates (ed.), *Self-Organizing Systems,* Plenum, 1982.

Hall of Fame

[1] For historical details of this crucial event, see Carl Benjamin Boyer, *The History of the Calculus, and its Conceptual Development,* Dover, 1959.
[2] *The Theory of Sound* (12), esp. Ch. 2.

Chapter 1

[1] See Zeeman (17), p. 4.

[2] For an elaborate and carefully considered alternate definition of *attractor,* see David Ruelle, Small random perturbations of dynamical systems and the definition of attractors, *Commun. Math. Phys. 82, 137-151 (1981).*

Chapter 2

[1] See Holmes and Moon, *J. Sound Vibr.* (1979)65(2), 275-296.
[2] An excellent history of this development is found in Cartwright, Nonlinear vibrations: a chapter in mathematical history, *Math. Gaz.* (1952)36, 80-88.
[3] For the history and more discussion of this model, see Rosen (13). And for the details of the mathematical analysis, see Hirsch and Smale (5).
[4] See Hirsch and Smale (5) for details.
[5] See H.I. Freedman, *Deterministic Mathematical Models in Population Ecology,* Decker, New York, 1980.

Chapter 3

[1] For a modern version of this example, see Hirsch and Smale (5) and Hayashi (3).
[2] See Hirsch and Smale (5), Ch. 10.
[3] See Rosen (13), Ch. 7, for more discussion of this scheme.
[4] See Schelleng, Scientific American, Jan. 1974.

Chapter 4

[1] *The Theory of Sound* (12), Art. 51.
[2] *The Theory of Sound* (12), Art. 42.
[3] If you haven't, this is a good time to begin. See Zeeman [17], Ch. 9, for seven applications of Duffing's cusp catastrophe to psychological behavior.
[4] An early study of harmonics in the Duffing ring is C.A. Ludeke, *Journ. Appl. Physics* 13 (1942), 215-233.

Chapter 5

[1] For historical details, see Cartwright, *op. cit.,* Also, see the outstanding text of the subject, Stoker (14).
[2] The Cartesian product construction is described in Volume 0 of this series.
[3] See Sacker, *Comm. Pure Appl. Math.* 18 (1965), 717-732.
[4] For full details of this response diagram, see Stoker (14).

BIBLIOGRAPHY

(1)   Abraham, Ralph H., and Marsden, Jerrold E., 1978.

Foundations of Mechanics, 2nd ed.; Benjamin/ Cummings, Reading, Mass.

(2)   Arnold, V. I., 1973.

Ordinary Differential Equations, (1978 english edition, tr. by R. A. Silverman); MIT, Cambridge (USA).

(3)   Hayashi, Chihiro, 1964.

Nonlinear Oscillations in Physical Systems; McGraw-Hill, New York.

(4)   Helleman, Robert H. G., (ed.), 1980.

Nonlinear Dynamics, (Annals, vol. 357); New York Academy of Sciences, New York.

(5)   Hirsch, Morris W. and Stephen Smale, 1974.

Differential Equations, Dynamical Sytems, and Linear Algebra; Academic, New York.

(6)   Hilton, Peter, (ed.), 1976.

Structural Stability, the Theory of Catastrophes, and Applications in Sciences, (Lecture Notes in Math., v. 525); Springer-Verlag, Berlin-Heidelberg-New York.

(7)   Hoppensteadt, Frank C., 1979.

Nonlinear Oscillations in Biology, (Lec. Appl. Math., vol. 17); Amer. Math. Soc., Providence, R.I.

(8)   Irwin, M.C., 1980.

Smooth Dynamical Systems; Academic, New York.

(9)   Markus, Lawrence, 1971.

Lectures in Differentiable Dynamics; Amer. Math. Soc., Providence, R.I.

(10) Nitecki, Z., 1971.

Differentiable Dynamics; MIT Press, Cambridge, Mass.

(11) Nitecki, Z. and C. Robinson (eds.), 1980.

Global Theory of Dynamical Systems, (Lecture Notes in Mathematics, vol. 819); Springer-Verlag, Berlin-Heidelberg-New York.

(12) Rayleigh, Baron, 1877.

The Theory of Sound, (2 vols., 1945 edition); Dover, New York.

(13) Rosen, Robert, 1970.

Dynamical System Theory in Biology; Wiley-Interscience, New York.

(14) Stoker, J. J., 1950.

Nonlinear Vibrations; Interscience, New York.

(15) Thom, Rene'

Structural Stability and Morphogenesis, (english edition of 1975, tr. by D. H. Fowler); Benjamin/Cummings, Reading, Mass.

(16) Winfree, Arthur T., 1980

The Geometry of Biological Time, (Biomath., v. 8); Springer-Verlag, Berlin-Heidelberg-New York.

(17) Zeeman, E. Christopher, 1977.

Catastrophe Theory; Addison-Wesley, Reading, Mass.

INDEX

Index numbers refer to panels, rather than pages.

Announcing the Visual Mathematics Library

— Film Series —

VISMATH LIBRARY FILM SERIES

The Lorenz System

by Bruce Stewart

The first in a series of high resolution computer animated movies on nonlinear dynamics, *The Lorenz System* shows a visual example of elementary chaos. Edward Lorenz's model of thermally driven convection is explained in a standard 16mm color film of 25 minutes duration. Because the subject itself is three-dimensional and dynamic, the film format can bring fundamental ideas from the research frontier within the reach of non-specialists.

The film introduces the fluid dynamical model leading to the dynamical system, and constructs phase portraits of the system for a wide range of parameter values. Ideas are introduced step by step, beginning with the notion of phase space itself. The presentation is entirely visual, without equations, but with frequent captions explaining the important ideas. The major bifurcations in the Lorenz system are seen, and the mainifold outstructure emanating from the equilibria is examined in the laminar, pre-chaotic, and chaotic regimes. The geomoetry of period doubling cascades is observed.

The Lorenz System is suitable for college-level students of differential equations, fluid mechanics, or nonlinear oscillations. Anyone who deals with nonlinear models of dynamics (in physics, chemistry, biology, ecology) can gain valuable insight from this film. *25 minutes, color, 16mm.*

VISMATH LIBRARY FILM SERIES

Chaotic Chemistry

by Robert Shaw, Jean-Claude Roux
and Harry Swinney

Experimental data from a stirred chemical reactor, plotted according to the graphical scheme of nonlinear dynamics, reveals a geometric figure essentially identical to the famous Rössler attractor of chaotic dynamics. (Section 3.4 of *Part Two, Dynamics: The Geometry of Behavior*). This film shows the experiment, the graphical scheme, and the structure of the reactor, in exquisite detail, using state-of-the-art computer graphics equipment. A brief written description accompanies the movie.
20 minutes, black & white, 16mm

VISMATH LIBRARY FILM SERIES

Chaotic Attractors of Driven Oscillators

by J.P. Crutchfield

This movie studies a series of classic nonlinear oscillators. The technique used is that of animated Poincaré sections. This is the temporal animation of cross sections through an attractor. A single Poincaré section is made by collecting the oscillator's position and velocity at a fixed phase of the driving force. The animation then plays back in time successive sections as the driving phase advances. The technique allows one to easily see and study the folding and stretching geometry around the attractors.

The movie presents five chaotic attractors taken from three different nonlinear oscillators. The first three examples come from Shaw's variant of the driven Van der Pol oscillator. (See Sec. 3.2 of *Part Two, Dynamics: The Geometry of Behavior*). The first exhibits the folding action of three "ears" on a torus attractor. The second "ribbon" attractor is the consequence of a period-doubling route to chaos. The final Van der Pol example reveals a complex attractor with visible fractal leaves. The movie illustrates the first attractor's symmetries by superimposing sections.

The fourth attractor comes from Duffing's oscillator (See Chapter 4, *Part One, Dynamics: The Geometry of Behavior*). With its thick fractal structure, it is reminiscent of oriental brush-stroked characters. The final example is the driven damped pendulum. As this attractor is spatially periodic and of infinite extent, a "five-well" segment is shown. With the animation it appears as a train of ocean waves continually breaking on a beach.

The movie *Chaotic Attractors of Driven Oscillators* was filmed during the fall of 1981 and premiered at Dynamics Day La Jolla, 4-6 January, La Jolla, California. A brief written description accompanies the movie.
12 minutes, black & white, 16mm

To Order

Please see enclosed Price List and complete the order form attached. Advance pay't required.

Introducing:

The Science Frontier Express Series

V O L U M E O N E
SCIENCE FRONTIER EXPRESS SERIES

The Dripping Faucet as a Model Chaotic System
by Robert Shaw

One of the most exciting recent developments in dynamical system theory has been the emergence of a better understanding of the **"chaotic transition,"** the change of behavior of many systems from periodic to nonperiodic behaviour. In this work, the author shows that the pattern of drops from a simple faucet makes such a transition, as the tap is slowly opened. This physical example is used to address the important general question: if a system is chaotic, how chaotic is it?

Information theory, the author argues, provides the appropriate tools for sorting out mixtures of determinism and chaos. Although this work describes the very latest results in the application of information theory to dynamical systems, the presentation is as nontechnical as possible. The text is illustrated by more than 60 pictures, and every effort has been made to make the material accessible to a wide audience. The result is a remarkably clear discussion of the concepts of "entropy" and "information" in the context of dynamical systems — easily readable, for example, by students of Shannon's book on information theory. The book ends on a more philosophical note, with a personal view of the issues which will loom largest in the future development of the subject.   _111 pages, 63 illustrations_

V O L U M E T W O
SCIENCE FRONTIER EXPRESS SERIES

On Morphodynamics
by Ralph Abraham

On Morphodynamics includes selected papers written by Dr. Ralph Abraham on models for pattern formation processes, morphogenesis, and self-organizing systems, showing the evolution of the **complex dynamical systems** concept over a fifteen year period. The works indicate a range of applications spanning the physical, biological, psychological, and social sciences.   _225 pp., 55 pp illustrations_

The volume includes:
1. _Stability of models_, 50 pp, 1967.
2. _Introduction to morphology_, 126 pp, 1972.
3. _Psychotronic vibrations_, 4 pp., 1973.
4. _Vibrations and the realization of form_, 18 pp., 1976.
5. _The macroscopy of resonance_, 8 pp., 1976.
6. _Simulation of cascades by videofeedback_, 5 pp., 1976.
7. _The function of mathematics in the evolution of the noosphere_, 15 pp., 1980.
8. _Dynamics and self-organization_, 28 pp., 1980.
9. _Dynamical models for thought_, 22 pp., 1981.

V O L U M E T H R E E
SCIENCE FRONTIER EXPRESS SERIES

Complex Dynamical Systems
by Ralph H. Abraham

Complex Dynamical Systems includes selected recent papers on complex models for physiological systems.   _125 pp., 43 pp. illustrations_

The volume includes:
1. _Categories of dynamical models_, 25 pp., 1983.
2. _Dynamical models for physiology_, 6 pp., 1983.
3. _Complex dynamical systems_, 5 pp., 1984.
4. _Chaos and intermittency in an endocrine system model_ (with H. Koçak and W.R. Smith), 41 pp., 1981.
5. _Orbital plots of dynamical processes_, (with A. Garfinkel), 19 pp., 1983.
6. _Cortisim_ (with A. Garfinkel), 11 pp, 1983.
7. _Endosim, a progress report_, 9 pp., 1984.

Order Form/Price List for:

AERIAL PRESS, INC.
P.O. Box 1360 - Santa Cruz, CA 95061
(408) 425-8619

(Eff. 10/84)

	Price	Ship/Handling
Dynamics: The Geometry of Behavior		
PART THREE: GLOBAL BEHAVIOR	$ 26.00	$ 2.00*

Pre-publication discount: orders placed before 12/31/84 receive 10% discount on price shown.

Part One: Periodic Behavior	32.00	2.00*
Part Two: Chaotic Behavior	26.00	2.00*
Disk Two: Chaotic Attractors in 3D	20.00	2.00*
Disk One: Periodic Attractors in the Plane	20.00	2.00*

*Prices are for shipping inside the U.S. only. For shipments outside the U.S., shipping costs are: Canada & Mexico: $4.00 for each item ordered. South America: $8.00 for each item ordered. Japan: order from Yurinsha, Ltd., Hongo, P.O.Box 63, Tokyo, 113-91, Japan. All other countries: Order from Birkhauser, Elisabethenstrasse 19, CH-4010, Basel, Switzerland.

SCIENCE FRONTIER EXPRESS SERIES

	Price	Ship/Handling
Volume One: The Dripping Faucet as a model Chaotic System by Robert Shaw	$ 15.00	$ 2.00**
Volume Two: On Morphodynamics, Selected Papers by Ralph H. Abraham	25.00	2.00**
Volume Three: Complex Dynamical Systems, Selected Papers by Ralph H. Abraham	15.00	2.00**

**Prices are for shipping inside the U.S. only. For shipments outside the U.S., shipping costs are: Canada & Mexico: $4.00 for each item ordered. South America: $8.00 for each item ordered. All other countries: $12.00 for each item ordered.

VISUAL MATHEMATICS LIBRARY FILM SERIES

	Price	Ship/Handling
The Lorenz System by Bruce Stewart	$190.00	$5.00***
Chaotic Chemistry by Robert Shaw, Jean-Claude Roux and Harry Swinney	190.00	5.00***
Chaotic Attractors of Driven Oscillators by J. P. Crutchfield	190.00	5.00***

***Prices are for shipping inside the U.S. only. For shipments outside the U.S., shipping costs are: Canada & Mexico: $6.00 for each item ordered. South America: $12.00 for each item ordered. All other countries: $16.00 for each item ordered.

PAYMENT MUST ACCOMPANY ORDER. PLEASE INCLUDE SHIPPING CHARGES.

Aerial Press, Inc.

P.O. Box 1360   Santa Cruz, CA 95060   (408) 425-8619

Name _____

Street _____

City, State, Zip _____

To charge to your VISA or MasterCard, please complete the following:

Card # _____   Exp. Date: _____

Signature _____   Phone #: _____

Item(s) Ordered	Price	Sh./Handl	Quantity	Total
1.				
2.				
3.				
4.				
5.				
6.				
7.				
8.				
California Residents add 6% sales tax				
(U.S. Dollars) TOTAL				

Payment Must Accompany Order.

Payment must be in U.S. Dollars. For orders outside the U.S., please send payment via either an International Money Order, or a check with American Banking Association (ABA) computer numbers, drawn on a U.S. bank.